The New Poetics of Climate Change

Environmental Cultures Series

Series Editors:
Greg Garrard, University of British Columbia, Canada
Richard Kerridge, Bath Spa University, UK

Editorial Board:
Franca Bellarsi, Université Libre de Bruxelles, Belgium
Mandy Bloomfield, Plymouth University, UK
Lily Chen, Shanghai Normal University, China
Christa Grewe-Volpp, University of Mannheim, Germany
Stephanie LeMenager, University of Oregon, USA
Timothy Morton, Rice University, USA
Pablo Mukherjee, University of Warwick, UK

Bloomsbury's *Environmental Cultures* series makes available to students and scholars at all levels the latest cutting-edge research on the diverse ways in which culture has responded to the age of environmental crisis. Publishing ambitious and innovative literary ecocriticism that crosses disciplines, national boundaries, and media, books in the series explore and test the challenges of ecocriticism to conventional forms of cultural study.

Titles available:
Bodies of Water, Astrida Neimanis
Cities and Wetlands, Rod Giblett
Ecocriticism and Italy, Serenella Iovino
Literature as Cultural Ecology, Hubert Zapf
Nerd Ecology, Anthony Lioi
The New Nature Writing, Jos Smith

Forthcoming titles:
Colonialism, Culture, Whales, Graham Huggan
Climate Crisis and the 21st-Century British Novel, Astrid Bracke
This Contentious Storm: An Ecocritical and Performance History of King Lear, Jennifer Hamilton

The New Poetics of Climate Change

Modernist Aesthetics for a Warming World

Matthew Griffiths

BLOOMSBURY ACADEMIC

LONDON • NEW YORK • OXFORD • NEW DELHI • SYDNEY

BLOOMSBURY ACADEMIC
Bloomsbury Publishing Plc
50 Bedford Square, London, WC1B 3DP, UK
1385 Broadway, New York, NY 10018, USA

BLOOMSBURY, BLOOMSBURY ACADEMIC and the Diana logo
are trademarks of Bloomsbury Publishing Plc

First published in Great Britain 2017
Paperback edition published 2019

Cover design: Paul Burgess/Burge Agency
Cover image © Shutterstock

A catalogue record for this book is available from the British Library.

ISBN: HB: 978-1-4742-8209-3
PB: 978-1-3500-9947-0
ePDF: 978-1-4742-8211-6
ePub: 978-1-4742-8210-9

Names: Griffiths, Matthew (Matthew J. R.) author.
Title: The new poetics of climate change : modernist aesthetics for a warming
world / Matthew Griffiths.
Description: London ; New York : Bloomsbury Academic, 2017. | Series:
Environmental cultures | Includes bibliographical references and index.
Identifiers: LCCN 2017000353| ISBN 9781474282093 (hb) | ISBN 9781474282109
(epub)
Subjects: LCSH: Climatic changes in literature. | Nature in literature. |
English poetry–20th century–History and criticism. | English
poetry–21st century–History and criticism. | American poetry–20th
century–History and criticism. | American poetry–21st century–History
and criticism.
Classification: LCC PN1065 .G75 2017 | DDC 809.1/936–dc23 LC record available
at https://lccn.loc.gov/2017000353

Series: Environmental Cultures

Typeset by Fakenham Prepress Solutions, Fakenham, Norfolk NR21 8NN

To find out more about our authors and books visit
www.bloomsbury.com and sign up for our newsletters.

For Ray and Isabel Griffiths: from myself to your shelves

Contents

List of Illustrations

Acknowledgements

This book began life as a doctoral thesis (although it did not end there), and I would like to acknowledge the support of those who helped me complete that undertaking. Ustinov College at Durham University provided several annual awards of a residential College Scholarship that enabled me to carry out my studies in the first place, and I am grateful to college staff and students for fostering such a welcoming, productive postgraduate community at Howlands Farm. Staff and students in Durham's Department of English Studies were likewise very supportive: I am grateful, in particular, to Dr John Clegg, who was the first to propose that I consider *The Anathemata* as part of this project, and also put many of my ideas to the test; to Dr Jack Baker, who offered valuable pointers on the work of Ezra Pound; and to Dr Jason Harding, whose thorough and challenging readings of my thesis, in his capacity as my secondary supervisor, enabled my argument to be more rigorous throughout. I also owe significant thanks to my supervisor Professor Timothy Clark, whose assiduous attention, sensitive direction and sympathetic support were essential throughout and beyond the researching and writing of this book. As my examiners, Professor Patricia Waugh of Durham University and Dr Harriet Tarlo of Sheffield Hallam University provided useful direction on how to develop the thesis. Meanwhile, Durham University Library – in particular, the Basil Bunting Poetry Archive in its Special Collections – the British Library and the Saison Poetry Library gave me the opportunity to consult a number of works in the course of my research. At Bloomsbury Academic, David Avital, Mark Richardson and Lucy Brown provided valuable help in steering the project home, and the copy-editing of Aruna Vasudevan tidied and tightened my academese. I must also thank the attentive Kim Storry at Fakenham Prepress, whose proofreading and patience have spared me plenty of typographical embarrassments. Needless to say, any remaining errors and omissions in the work are my own.

Some of my research has been presented in earlier versions at several events, among which I must give particular credit to the committee of the

Ustinov Seminar, Durham University and to the organizers of the Culture and Climate Change symposium at Bath Spa University in summer 2010, at each of which I presented research on *The Waste Land*, as well to the convenors of Durham University's Inventions of the Text seminar series and the organizers of the University of Exeter's Environment and Identity Conference in summer 2011, which both saw me present on contemporary climate change poetry. More generally, I have benefitted from being a member of the Association for the Study of Literature and the Environment (UK & Ireland), which has provided a supportive network of like-minded academics, with whom I have learnt and discussed much about ecocriticism, and given me the opportunity to approach series editors Greg Garrard and Richard Kerridge with the proposal for this book, which they kindly endorsed. I was very glad, too, to receive the encouraging responses of the proposal's four anonymous readers, and in particular the incisive and sympathetic comments of the manuscript's anonymous reviewer, which have significantly helped me refine and hone my argument.

With a book of this nature, I am dependent on a number of publishers, who have generously extended permission to quote from the work of the poets I have considered. 'The Sorcerer's Mirror' by Andrew Motion is copyright © Andrew Motion, 2009, used by permission of The Wylie Agency (UK) Limited. My thanks go to Frances Presley and to Tony Frazer of Shearsman Books to let me quote from 'Triscombe stone'; Faber & Faber Ltd to allow me to quote from *The Poems of T. S. Eliot*, Eliot's *Selected Prose* and *On Poetry and Poets*, from Wallace Stevens's *Collected Poems* and from David Jones's *The Anathemata*, as well as to quote from Jo Shapcott's *Of Mutability* and Ezra Pound's *Selected Poems and Translations*; Carcanet Press Limited, Manchester, UK, for permission to reprint copyrighted material from Jorie Graham's *Sea Change* and from the *Collected Poems* of William Carlos Williams; New Directions Books for permission to allow me to quote from *Selected Poems and Translations of Ezra Pound* and *Literary Essays of Ezra Pound* and from *The Collected Poems of William Carlos Williams*; Bloodaxe for permission to quote from the *Complete Poems of Basil Bunting* (in the Faber edition) and *Briggflatts*, and to reproduce Bunting's diagram of the latter's structure, as well as allowing me to quote from Fleur Adcock's *Poems 1960–2000* and Peter Reading's *–273.15*; the Trustees of the David Jones Estate for allowing me quote from his work; while 'shut the fuck up and drink your gin & tonic' by D. A. Powell is quoted with the permission of The Permissions Company, Inc., on behalf of Graywolf Press, Minneapolis, Minnesota. I am likewise grateful to the Quantock Hills Area of Outstanding Natural Beauty Service for supplying an image of Triscombe Stone, Mario Petrucci for permission to reproduce material for an interview I conducted with him in 2009, with additional thanks for his support and encouraging remarks; and to the Special Collections Research Center, University of Chicago Library for allowing me to reproduce material from Basil Bunting's correspondence with *Poetry* magazine. Excerpts from T. S. Eliot's *Collected Poems 1909–1962* are also copyright 1936, renewed 1964, Houghton Mifflin Harcourt Publishing Company; excerpts from *The Waste Land: A Facsimile and Transcript of the Original Drafts including the Annotations of Ezra Pound* copyright 1971 Valerie Eliot; and excerpts from *The Selected Prose of T. S. Eliot* are copyright 1975 Valerie Eliot. All excerpts from these are reprinted by permission of Houghton Mifflin Harcourt Publishing Company, all rights reserved. Quotations from 'The Man on the Dump', 'The Idea of Order at Key West', 'The Snow Man', 'The Planet on the Table', 'The Plain Sense of Things', 'A Postcard From the Volcano', 'Notes Toward a Supreme Fiction', 'Credences of Summer', 'Sea Surface Full of Clouds', 'The Bouquet', 'The Poems of Our Climate', 'Variations on a Summer Day' and '*Poésie Abrutie*' are from *The Collected Poems of Wallace Stevens* by Wallace Stevens © 1954 by Wallace Stevens and copyright renewed 1982 by Holly Stevens; excerpt(s) from *The Necessary Angel: Essays on Reality and the Imagination* by Wallace Stevens are copyright © 1942, 1944, 1947, 1948, 1949, 1951 by Wallace Stevens; and excerpt(s) from *Opus Posthumous: Poems, Plays, Prose* by Wallace Stevens © 1989 by Holly Stevens and © 1957 by Elsie Stevens and Holly Stevens, copyright renewed 1985 by Holly Stevens; all of these are used by permission of Alfred A. Knopf, an imprint of the Knopf Doubleday Publishing Group, a division of Penguin Random House LLC. All rights reserved.

Finally, I am very grateful to my mother and father, Isabel and Raymond Griffiths, who provided moral and financial support during the course of my doctoral research and, over many years, fostered in me the intellectual aspiration to undertake such a project. Thanks, at last, to Allison Siegenthaler for much love, support, encouragement and patience, without which the work would never have reached this stage.

Climate Changes Everything

The climate change poem

In 2009, as his term as UK poet laureate ended, Sir Andrew Motion was asked to write a piece that would feature in the *Guardian*, among commissions from other writers, 'To support the launch of the 10:10 campaign to reduce carbon emissions' (Motion 2009b). The poem was to be set to music by the late Sir Peter Maxwell Davies for the University of Cambridge. Motion composed a five-sonnet sequence entitled 'The Sorcerer's Mirror'.

The first sonnet begins with the narrator explicitly locating himself in time and place: 'Midnight and midsummer in London. / I ... stand in my quarter-acre of garden' (ibid.). The end-stopped opening line is still and verbless, fixing a moment in time. As the poem progresses, the narrator acknowledges that he cannot remain in his pastoral vantage point, a small ('quarter-acre') green space in the encroaching metropolis. The solace he seeks is reminiscent of that sought by the narrator of Andrew Marvell's 'The Garden'; yet Motion's choice of phrasing in 'at my back the spacious mulberry tree' (ibid.) more clearly alludes to the earlier poet's 'To His Coy Mistress', whose narrator declares 'at my back, I always hear / Time's wingèd chariot hurrying near' (Marvell 2005: 51, lines 21–2). Motion's allusion therefore sets up a tension between stillness and change, but one, ironically, in which the present offers greater stasis than the poetic tradition. For instance, whether we interpret his phrase 'what passes for its [the earth's] sleep' (Motion 2009b) as referring to the busyness of nocturnal London or to the continuation of natural processes while the city's inhabitants are in bed – or, indeed, as a metapoetic recognition that likening night to human sleep is a fictive, anthropomorphic device – it emphasizes the provisional quality of the calm he originally invokes. Yet the narrator strives to remain in the green eye of an urban storm, and signals a separation between himself and nature at the end of the first sonnet: 'the dark earth wakes and I look on', he remarks (ibid.), putting himself in the position of a privileged spectator.

The strategic advantage of creating a lyric persona and an everyday environment, as Motion does here, is to engage the reader in familiar, shared experience. However, introducing climate change then presents a challenge. Where and how do we experience it in this domestic milieu? Motion progresses outwards from his seclusion by attending to 'the sour music of traffic cruising close' (ibid.), a banal juxtaposition of solitude with engine noise, and of the (relatively) natural garden with the automobile as a totem of pollution. The cars are at least kept at a distance – although 'close' they are outside the garden. To cover any further ground, however, the narrator has to be 'swept on a breeze / which was … pure and simple once', and which 'carries and scatters [him] / over the polar cap' (ibid.). Again, he attempts to put distance between the lyric clarity of this vision and the pollution he witnesses, although the distinction is not now geographical but temporal, and thus nostalgic – the breeze was 'pure and simple *once*' (my italics). A vision of Nature,[1] in the common usage that signifies the wild or green world, is associated with the narrator's sense of self in his garden in the first sonnet, and then in the second, with some unspecified point in the past when the wind was unpolluted. Already Motion's Romantic leanings are becoming clear.

Now apparently scattered, his self still commands an integrated voice, and the narrator remains a distanced observer who 'looks on' as he did in the first sonnet. A tone of pastoral retreat and pleasure informs both the view of green nature in the garden, and the polar ice: 'every luminous, upside-down meadow stitched / with gorgeous frost-flowers and icicle grass' (ibid.). By assimilating this imagery to the pastoral tradition, Motion makes it doubly familiar, because we already recognize the 'snapped-off sea-ice' and the 'rising tides / overflowing their slack estuaries and river basins' later in the sequence (ibid.) as tropes of climate change thanks to three decades of news reports and natural history documentaries. Motion's reference to 'the already famously lonely polar bear' (ibid.) is half-hearted, then, both as a recognition in itself, and as an enactment of the compassion fatigue engendered by the visuals of news media.

Motion's vision of an untouched polar region is elaborated in the second sonnet: 'three thousand years have worked through / and sculpted [it] in silence' (ibid.). The ice here represents a work of art three millennia in the making, one which the present state of affairs puts in jeopardy – a conceit that supposes a preceding continuity in the order of nature against which contemporary climate change can be distinguished. The invocation of 3,000 years, which accounts for a substantial part of the history of civilization, gives this vision some weight. However, it has to be scaled to the time frame of human existence to do so. In *A Cultural History of Climate*, Wolfgang Behringer notes that 'there has been no permanent ice during 95 per cent of the earth's history. Statistically, warm periods are the characteristic climate of our planet' (2010: 20). Motion's 'three thousand years' is a much shorter period of time, corresponding roughly with Behringer's observation that 'The long warm and dry period of the Bronze Age gave way around 800 BC – roughly 2,800 years ago – to [a] cooler climate' (ibid.: 58). Nevertheless, Motion tacitly valorizes the formation of ice by likening it to a work of art ('sculpted') and hinting that he might find there the seclusion ('silence') that he is unable to locate in London. In this aesthetic metaphor, the apparent longevity of arctic ice is given human value; but without being so prized, it has no real claim on permanence, given the much longer timescales described by Behringer of which it is a part. Motion does, admittedly, attempt to shift this human focus, but only towards the end of the fifth sonnet, where he recognizes that it is impossible to enclose domestic space apart from the rest of the world: he moves 'quickly over the threshold' back into the house, but 'one look is enough to show the bare horizon behind' (2009b). The door between the human province and the world is open and cannot be completely shut. Nevertheless, this still maintains a distance between himself and the 'bare horizon', which replicates the earlier distance between London and the pole, suggesting that climate change is more accessible at the latter location than the former.[2]

In his polar vision, Motion also seeks to stabilize a vision of Nature against which climate change might be measured; he indulges in something that the critic Lawrence Buell is alive to in our experience of place, 'to fantasize that a pristine-looking landscape seen for the first time is so in fact' (2001: 68). But as the concept of Nature becomes more fixed and certain in such a vision, so climate change also crystallizes into a particular entity, reduced to one of the topics that constitute the media category of environment. Motion is self-conscious enough about borrowing from the climate discourse of other media, such as 'the already famously lonely polar bear' (2009b).

Motion's poem raises some of the key issues surrounding poetic engagements, or attempts at engagement, with climate change. For instance, his role as poet laureate indicates the status of climate change as a public issue, the commission emphasizing the poet's role as bearing witness. In an interview with Richard Eden in the *Telegraph*, Motion alludes to this function, declaring: 'To me, climate change is so bleeding obvious. Anyone who thinks it's not happening should get outside more' (2009a). If that is the case, why do we need a poem to tell us as much? At the same time, the narrator of 'The Sorcerer's Mirror' only advances as far as his back garden; it takes an imaginative projection to the pole to witness climate change happening. Contrary to Motion's claim, the poem implies that 'getting outside' would only work if we could go to the calving face of the ice itself. This tension attests to the difficulty of bearing witness to a range of phenomena[3] whose complexity is largely not amenable to sense experience. The strain of trying to reconcile local and polar is further complicated by Motion's choice of mode. I have already suggested that the sequence begins in the pastoral vein that Marvell is working in, and draws straightforward contrasts between natural purity and current trends. His *Telegraph* interview makes the choice of mode explicit: 'I've written a lament about it [climate change] which has the air of a call to arms' (ibid.). The movement from 'lament' to 'a call to arms' figures a process whereby poetic elegy is designed to inspire a presumed reader or audience politically, and this surrounds the sequence with further interpretative frames.

'The Sorcerer's Mirror' enables me, then, to pose some key questions about climate change poetry. What, for example, are we 'lamenting' as the climate changes? Can a poem, whether or not it is explicitly about climate change, spur us into political action? If by changing our intentions we could prevent further climate change, would that suggest that the climate is already changing only as a result of our conscious, intentional actions? Does such action depend on personal epiphanies such as the narrator's being 'reflected back at [him]self, crouched like a guilty thing' in his French windows (2009b)?

These invite further, broader questions. How is climate change constituted as a media topic, a political issue or an atmospheric and oceanic phenomenon?

KEY QUESTIONS ABOUT CLIMATE CHANGE POETRY

Can it consist in the opposition of certain tropes of environmentalist invective like 'the sour music of traffic' and 'the miserable sky-litter / of planes circling in their stack' (ibid.) with totems such as polar bears and icecaps? Must poetry about climate change belong in the tradition of the pastoral or the elegy? Does it demand a moralistic tone? What do these formal considerations of the poetry suggest about climate change, and our engagement with it? And what alternative models or approaches might there be?

As a topic of discourse, climate change already attracts a plethora of concerns according to the frame we unconsciously or intentionally position around it. Poetry as a cultural practice is therefore subject to a broader tendency in which climate change is attached to preconceived theoretical or ideological frameworks. The geographer Mike Hulme remarks that the 'quality of plasticity' that the idea of climate changes possesses allows it 'easily to be appropriated in support of a wide range of ideological projects' (2009: xxviii). Rather than being seen as necessitating fresh engagement with the phenomenal world, climate change has too often been taken as a reason to retrench and confirm cherished, nostalgic views of natural stability. While literature is in a position to interrogate culturally given representations through formal innovation, Motion is by contrast too ready to contain a narrative of climate change within a conventional, narrative form[4] rather than use his position in the tradition to challenge or advance that tradition or articulate a fresh or engaging conception of climate change.

THE POINT OF POETICS TO ARTICULATE FRESH CONCEPTIONS

Yet Motion's is not the only approach to be taken by poets in response to the phenomena and politics of climate change, and alternative strategies have developed alongside scientific and societal understanding over the past few decades. The form, focus and style of this verse – its poetics – shapes its understanding of climate change, and as conventional strategies such as Motion's demonstrate limited insight into the phenomena, lagging behind those of other media, the emergence of alternative aesthetics signifies the search for more nuanced and sophisticated engagement with the phenomena. Just as climate change disrupts and disturbs pastoral or Romantic conceptions of nature, so formally experimental, avant-garde writing in the Modernist tradition can offer the opportunity to rethink rather than reinforce conventional reader–writer relations.

THE REASON FOR EXPERIMENTAL AVANT GARDE WRITING IN THE MODERNIST TRADITION

Frances Presley's 'Triscombe stone', also collected in 2009, makes a fuller engagement than Motion does with the problematic aspects of climate change

COMPARISON WITH PRESLEY'S [*TRISCOMBE STONE*]

LOCALISM

I have raised above. While Presley's poem refers to an actual site – Triscombe Stone in Somerset, UK (see Figure 1) – her localism is not the jumping-off point for an imaginative projection, as Motion's garden is, but rather a focus for all that cannot be placed in time or space. Indeed, her poem opens with a question about place, 'Which wood are we in?', an enquiry repeated when the first answer, 'Maritime sessile oak', seems insufficient: 'No, which wood are we in? I thought it was scrub oak' (Presley 2009: 47). This demonstrates a disjunction between an apparent pair of voices, with the question inviting a geographical reply but receiving a taxonomic one; even the simplest experience yields to multiple interpretations, and the necessity of expressing this entails a dialogue, rather than the kind of lyric voice with which Motion opens his poem.

DIALOGUE FORMAL DEVICE

This formal device enables Presley to set the titular 'stone' in a broader context of both time and space without the magical projection to which Motion has to resort to maintain a lyric identity and a coherent narrative. It is through a broadening in the variety of voices that the present of Presley's poem begins to intersect with both human history and climatic time: 'It is thought to be from the Bronze Age. The ancient of days lowers himself onto its cold stone which provides a convenient seat' (ibid.). This offers a concise example of the diversity of discourses that become attached to climate change – in this case, popular science and self-consciously poetical idioms represent different timeframes that climate change juxtaposes.

The image of the ancient stone 'half fixed in concrete' (ibid.) also signifies in several ways. Although we may be invited to contrast stone and concrete – that is, distinguish between nature and culture – the poem suggests that the stone warrants attention as 'rock art' (ibid.). Both concrete and worked stone are therefore products of human artifice, but by their juxtaposition they elongate our sense of civilization at least as far back as the Bronze Age, around 3,000–4,000 years ago. The continuity is emphasized by the placement of one of the voices in time 'In 1991', though the civilization of the present shares its contingency with that of the past: 'She says that only two degrees of warmth makes such a difference, and that was true in the Bronze Age' (ibid.). That recognition of contingency is also seen in one voice's 'discover[ing] for the first time that we only had eighty years left', forecasting an end to time as it is humanly conceived and measured ('*we only had eighty years left*').

Figure 1 Triscombe Stone, Somerset, UK © Quantock Hills AONB Service.

Presley's switching between perspectives engages with the questions of scale that climate change presents, in contrast to Motion's easy invocation of three millennia in imagining the arctic sublime.

In addition to locating the present in the context of history, Presley's poem situates the local within the global. By reference to the very category of 'rock art' and the observation 'that there is a pattern to it' (ibid.), Triscombe Stone is recognized as being part of more widespread cultural practices. That 'pattern' is only observable with dedicated research during the present, though, and those who originally placed the stone are unlikely to have been conscious of it, at least to the extent that we can be today. Presley demonstrates the difficulties of assuming our immediate experience offers a stable truth, given that the significance of the stone has changed over time; for whatever purpose it was originally erected, it now requires 'concrete' support, while it has come to be regarded as 'a hazard to cars'; both of these reveal how intention is not the only factor in determining the agency of entities such as 'rock art' (ibid.) that are produced by both natural and human processes. The significance of language is likewise subject to the process of change through time, as is evident in the direction of our attention in the lines 'recent wet ground conditions that have caused roadside damage by Dead Woman's Ditch' (ibid.). Here, the focus is on the present inconvenience, 'roadside damage', and the evocative name of the ditch merely serves as a setting even as it points up the distinction between a feminized landscape and the masculine imposition of a road over it.

Rather than exhibiting an overt distaste for cars, as Motion does in the line 'sour music of traffic' (2009b), Presley's lines here show both the everyday presence of the vehicles and their vulnerability to natural 'hazards' and 'damage'; this instances the fragility of human artifice, while also recognizing that we are party to environmentally deleterious practices, even in a context where '*There is proof of global warming at last*' (Presley 2009: 47; author's italics). While 'at last' suggests that the voice is conscious of global warming ahead of this 'proof', that awareness needs, like the Triscombe Stone, to be set as though in concrete by the very process of finding proof, hence the relief the phrase expresses at the close of the italicized sentence. That process is an explicitly mediated one as well, revealing the significant role language, in particular the discourse of the press, plays in our comprehension of the phenomena: 'I was forced to scan five newspapers', Presley writes, in order to discover this 'proof' (ibid.). This mediation is also seen in the way she transplants into literary form two justified columns of text, of the kind found in news reports such as 'the pink columns of the FT' that 'provided the most accurate and factual information, and were most likely to provide source references' (ibid.). Sense and form both demonstrate an awareness that climate change is always already a cultural production in a much more effective way than Motion's abashed reference to the 'already famously lonely polar bear' (2009b).

The process of representation is foregrounded towards the end of the poem, which ostensibly concerns the retouching of photographs: 'I can then clean the document, decrease the colour depth and optimise the image' (Presley 2009: 47). The last term, 'image', has a particular resonance in the context of poetry, suggesting that we must remain conscious of the process of reproducing the natural world in visual and literary art alike, never pretending that we can achieve immediate access to the complexity of phenomena in their entirety – a particularly salient point in the context of climate change.

That uncertainty can also be seen in the way that the poem not only opens with a question, but closes with another, choosing to resist a resolution and opening it up into a world of possible responses. The last sentence begins with speculation on the past that has been evoked, 'The long highways, who travelled', but in the transition to 'and why are you walking below me through watery lane?' (ibid.), it is unclear whether this speaker is addressing an imagined traveller of the past, or a companion in the present with its wet ground conditions. The ambiguity of perspective here not only disrupts our sense of perspective – is the past or the poem's present being questioned? or is it an address to the reader in the future? – but the seeming high ground of the poem images the difficulty of asserting an exterior, moral position, indicating that we are all implicated in climate change.

That position contrasts with Motion's in 'The Sorcerer's Mirror', but the difficulties he encounters in adopting a lyric focus for writing about climate change suggest that its phenomena are not so easily co-opted to the literary tradition, or at least to the traditions in which Motion chooses to position his sequence. Where his sequence is indicative of lyrical, proto-Romantic readings of climate change, Presley's poem shows the potential for more disjunctive, ambiguous and associative writing. The final question of her poem may gesture at lines from 'What the Thunder Said', the final section of T. S. Eliot's *The Waste Land*, 'Who is the third who walks always beside you?', 'But who is that on the other side of you?' (2015 [1922]: vol. I, 69, lines 359, 365) to hint at a post-apocalyptic wilderness, albeit an inundated one rather than Eliot's parched desert.

Presley's experimental approach, responding to climate change rather than seeking to fit it to pre-thought forms and modes, is part of the innovation necessary to poetry; it fulfils part of what Eliot in his own criticism entitles 'The Social Function of Poetry', in which 'there is always the communication of some new experience, or some fresh understanding of the familiar, or the expression of something we have experienced but have no words for, which enlarges our conscience or refines our sensibility' (1957: 18). Because the understanding expressed in 'The Sorcerer's Mirror' has in contrast lagged behind both scientific and popular discourse on the matter, it has not led the innovation in language that would take the complexities of the changing climate into account. In both its thought and practice, I propose, Modernist aesthetics is much better equipped to articulate the complexities and nuances with which climate change confronts us than is writing such as Motion's that appropriates – or rather, misappropriates – traditions of nature poetry to do so.

Criticism and climate change

The value of reading climate change through Modernism, and indeed Modernism through climate change, is that both disrupt previously cherished conceptions of the world; the fact that our imagination of Nature is only a part of what is disrupted by the emergence of both strengthens the case for such readings, because both Modernism and climate change force us in different ways to recognize that the cultural and the natural are always already entangled, and that we cannot make the latter our only focus. Further evidence for the value of considering climate change as a characteristically Modernist problem can be seen in the fact that both have, until recently, been neglected in the attentions of environmentally inflected literary criticism, or ecocriticism. This shared neglect is not arbitrary – it is precisely because both Modernism and climate change have ostensibly little to do with uninterrogated visions of Nature that they have been overlooked.

For instance, having declared that 'Modernism was never very green—if by green we mean an astute awareness of biodiversity, vigilance against pollution and overdevelopment, care for bioregional conservation, and an earth-focused activism that goes beyond human-centred interests', Joshua Schuster goes on in *The Ecology of Modernism* to explain succinctly that 'there are few ecocritical studies of the larger legacy of modernism, especially compared to the waves of attention shown to the ecological implications of British and American romanticism, and to global environmentalist writing since the 1960s' (2015: 3, 7). Similarly, in her survey of 'Ecocriticism and Modernism' for *The Oxford Handbook of Ecocriticism*, Anne Raine observes that 'modernism defines itself in opposition both to nature itself and to the two literary genres—realist prose and romantic nature poetry—that ecocritics tend to champion' (2014: 99), and hence has received little scrutiny by such critics.

Similar terms are used by Adam Trexler in his account of ecocriticism's foundational shortcomings when it comes to climate change: 'The field's preoccupation with life sciences led ecocritics to diagnose climate change as a human incursion into ecosystems or [into] Nature writ large, rather than a process that inextricably binds together human and nonhuman systems. As a result, ecocritics did surprisingly little to analyze anthropogenic global warming for the first decade and a half of the movement's existence' (2015: 17). Climate change represents a novel category of problem, not readily fitting into culturally predominant understandings of Nature, particularly because it compromises the realms conventionally inscribed as 'human' and 'natural'.

So, even though the discipline of ecocriticism emerged in tandem with a growing awareness of climate change during the late 1980s and early 1990s, ecocritics rarely referred to the phenomena, preferring instead to establish a canonical tradition that would justify reading the environment at all. In doing so, they consciously modelled their project on those of feminist, postcolonial and Marxist criticism in that they sought to reclaim marginalized writers and overlooked genres such as nature and, later, environmentalist writing for the canon, as well as bestowing fresh value on the Romantic and Transcendentalist corpus.

As far as ecocritics are concerned, climate change has tended to remain a challenge to thought, so to engage with its complexity in literary theory and practice that challenge must be addressed rather than evaded. This book will attempt to articulate a range of problems that climate change poses, and, by rereading Modernist texts with these in mind, propose some potential responses. It is for its challenging, and sometimes rejection, of Romantic paradigms that Modernism can help us accelerate a break with their influence on the critical and, more importantly, popular imagination of Nature; the work of T. S. Eliot represents a forthright repudiation of the Romantic, while there are more nuanced modifications of it in the poetry of Wallace Stevens, Basil Bunting and David Jones, the poets to whom I subsequently refer in establishing a new poetics for climate change. In the remainder of this chapter, though, I intend to articulate the key problems that climate change presents for conventional understandings of nature and even the environment, in particular in literary critical approaches. This will prepare us for subsequent exploration of Modernist poetics as an alternative way of reading such complexity in a nuanced and more sophisticated fashion.

Greens see things in waves

Where the first wave of ecocritics do attend to climate change, the way in which they read it in the Romantic paradigm is instructive because it helps to illuminate why the phenomena are so difficult to comprehend through such traditions. A brief look at the problems early ecocriticism faced in doing so, and how subsequent critics and thinkers have responded to these, will help me articulate some of the broader issues that this book will address, before I go on to explain why Modernist thought and poetics are particularly well positioned to enable us to do so.

In his landmark work *The Environmental Imagination* (1995), Lawrence Buell makes a pioneering case for the valorization of the natural environment through literature, in particular Thoreau's *Walden* and the canon of American writing to which it belongs. Buell argues that the work of environmental criticism is not just a parallel to but a radical development of the examination of cultural difference through literature that has been undertaken by postcolonial or feminist critics. In comparison to these agendas, he claims, 'by far the single most significant aspect of cultural difference with which we shall have to reckon pertains neither to ethnicity nor to gender but to anthropocentrism' (Buell 1995: 20), that is, the practice of making humans the focal point of writing and thought and placing the nonhuman in a secondary role.

Despite this stated intention, however, Buell finds it very difficult to relinquish human concerns in his critical approach. By choosing to focus on Thoreau, he already makes a conscious investment in the authority of selfhood as it has been conceived in the tradition of nonfiction nature writing, whereby the author serves as a reliable, disinterested observer for the reader. When Buell does then offer an aside on (human responsibility for) climate change, he therefore manages to exclude it effectively from critical consideration, because his attention is directed by what he considers possible in literature: 'the psychic health of an individual in a relatively self-contained subculture, and the health of that subculture as a whole, can be altered more easily than the rate of global CO_2 emissions', he writes (ibid.: 295). Buell's remark centres the work of his brand of ecocriticism in the self, disavowing the possibility of engaging with phenomena on a different scale.

Similarly, while Buell devotes considerable discussion to nature's significance in literature, his understanding of it refines rather than problematizes its definition, doubling down on popular conceptions in the American tradition. His vision for instance depends on 'seasonality' as a 'bedrock' for nature, which he describes as 'in some sense an obstinate objective given' (ibid.: 242). Bear in mind that seasonality, while not a function of human existence on earth, is roughly contemporary with it, representing the conditions obtaining in the current interglacial – conditions that have also enabled the emergence of civilization.[5] To conceive of these conditions as timeless is to ignore their contingency, and if we continue to imagine them in this way, as Motion does, then it becomes much more difficult to recognize why and how climate change has emerged.

In contrast to Buell's approach, global warming is the starting point for British critic Jonathan Bate's *Romantic Ecology* (1991), and he considers it in the light of its political context.[6] The 'leading stories on the evening television news' are before him, or at least in his mind's eye, as he writes, and these concern events in the USSR and newly unified Germany, as well as research 'that there are links between freak weather conditions and global warming' (Bate 1991: 1); the anecdote combines observation of the changing climate with politics and the mechanisms by which we bring nature into culture, via broadcast media and public debate, as we have seen in Presley's poem. Unlike Buell, Bate draws the warming globe further into his literary-critical method because he sees that its potential to disrupt seasonal procession could complicate future readings of the Romantics: Keats's ode 'To Autumn' is, he says, 'predicated upon the certainty of the following spring's return; the poem will look very different if there is soon an autumn when "gathering swallows twitter in the skies" for the last time' (ibid.: 2; citing Keats 2001 [1819]: 325, line 33). Bate thereby opens up the possibility of considering a text in undetermined futures. Indeed, in writing that 'One effect of global warming will be (is already?) a powerful increase in the severity of winds in northern Europe' (ibid.: 2), he is also alive to the possibility of an undetermined present, as his hesitancy signals that we cannot gain ready perspective on an environmental shift we might already inhabit.

Bate warms to his theme in his later piece 'Living with the Weather', where he takes Marxist, historicist readings of the tradition as a starting point for

reconsidering 'the legacy of romanticism in our age of eco-crisis' (1996: 435).
His key observation is that, whereas 'the scholar's elucidation of sources' (ibid.:
432) highlights allusion and influence in a text, and a politically oriented critic
recovers the 'human agency' surrounding the poem (ibid.: 437), the ecocritic
can distinctively draw our attention to environmental conditions at the time
of composition. Bate thus restores agency to nonhuman forces that have been
closed down in both literary and political analysis, because traditionally, the
'constancy of nature was something against which to measure the vicissitudes
of culture' (ibid.: 439); Bate's work opens for ecocriticism the possibility of
mapping nonhuman agency. The traditional critical tendency to take nature
as read and concentrate instead on human agency – criticism's anthropocentrism – inevitably overlooks crucial contexts, particularly those which
evince that 'nature is not stable' (ibid.). Nevertheless, this approach doesn't
validate Bate's claim that 'Global Warming Criticism is about to be born' (ibid.:
436),[7] because it addresses historical environmental conditions, rather than
the present-day implications of 'global warming'. Climate change then is used
more of a prompt to investigate previous environmental contexts, rather than
rethink them.

In reforming literary criticism to address the emergent environmental
crisis, both Buell and Bate turn to traditions that offer sublime, if problematic,
visions of the natural world: respectively, American Transcendentalism and
British Romanticism. In so doing, they try to accommodate crisis within
historic texts that accord with a particular conception of nature as a subject of
representation, rather than considering alternative visions of the nonhuman
world. But if climate change can affect a poem such as 'To Autumn' written
before the phenomena were even understood, why limit our retrospective
reading of it to the Romantic tradition? Climate change can, potentially,
colour our reading of any text of the past: we might make productive
ecocritical insights, and insights about ecocriticism, when reading texts that
resist a given vision of nature. Even when Bate reads outside that tradition,
he does so to reaffirm its influence, and those Modernists who merit consideration in his *The Song of the Earth* (2000), Wallace Stevens and Basil Bunting,
seem to do so not for their poetics but for their post-Romantic concern with
nature as a subject. The critic Timothy Morton offers a valuable question when
he proposes: 'The time should come when we ask of any text, "What does this

say about the environment?" In the current situation we have already decided which texts we will be asking' (2007: 5). Our readings should emphasize the adaptability of the literary to changing climates, whether literal or figurative, so that we regard literature as much as a series of dynamic processes as the environment is.

When ecocriticism does move into a second wave, it transitions between what Patrick D. Murphy describes as nature writing and environmental writing: the latter, unlike the former, 'does not stop at describing the natural history of an area, but instead, or in addition, discusses the ways in which … forms of human intervention have altered the land and the environment', he writes (2000: 5). Nevertheless, even while considering a much wider body of work than the Western canon, the second wave limits engagement with environmental crisis by charging ecocriticism with reading texts that confirm its interest in the topic – the 'Toxic Discourse' Buell describes in *Writing for an Endangered World* (2001: 30ff.), for instance – rather than examining the significant implications of anthropogenic environmental change for literary theory and practice. After all, considering how culture deals with climate change and considering how climate change deals with culture are distinct intellectual activities.

Murphy at least recognizes the limitations in only reading texts that already deal with environmental crisis, and seeks to give the discipline sufficient tools for it to look farther afield. Like Morton, he asserts that a 'reader has a right to expect that a general critical orientation would be applicable, at least to some extent, to every literary work' (ibid.: 16); hence we might make productive ecocritical insights, and insights about ecocriticism, when reading texts that resist a given vision of nature. He maintains that a wider scope can help 'exemplify how departures from Enlightenment realism' can be valuable, if only for the ways that they 'can intensify the themes found in environmental literature' (ibid.: 181), and it is for this reason that reading Modernism offers productive opportunities for denaturing ecocriticism.

If criticism is to be informed by climate change, our critical practice must be as pervasive and connective as the phenomena of climate change themselves: 'We can't rigidly specify anything as irrelevant', Morton points out later (2010a: 30). Reading texts for their representations of Nature, then, is as self-confirmatory as it is to read environmental crisis in the texts of environmental crisis, because Nature's givens only represent the environment that humanity has experienced in its time on the planet. For example, the positioning of human time frames alongside climatological ones that Motion is obliged to attempt in 'The Sorcerer's Mirror' offers a case in point. Behringer observes 'Only during the Holocene', beginning around 10,000 years ago, 'did the environment we now think of as "natural" make its first appearance', and the global warming that then occurred 'is associated with a fundamental shift to culture with more diverse and sophisticated features than before' (2010: 42–3). This means that our imagination of Nature depends on a particular interval in geological time, and is not as timeless as we might expect, without such 'bedrocks' as Buell imagines. In fact, Nature merely represents the conditions of our current interglacial episode, from about 10,000 or 13,000 years ago, as Behringer suggests.[8] Under such circumstances, we need to consider how to understand nature in the light of climate change.

The nature of nature

One of the key problems with the ideology behind first-, and to a lesser extent, second-wave ecocriticism is identified by the sociologist Bruno Latour in his book *Politics of Nature*. He cautions that green activists 'have come up with nothing better than a nature already composed, already totalized, already instituted to neutralize politics' (2004: 3). This is evident, for example, in the way Motion envisages that, by juxtaposing a vision of the polar sublime bluntly with urban living, any objections to taking action on climate change will be nullified. In contrast, Schuster writes that 'Modern ecology, and a critique of modernist ecopoetics, begins when one is no longer sure what nature means, when one does not begin with preset assumptions of what nature should be when nature does not have ontological, aesthetic or moral certainty and is an open, urgent question' (2015: 21). Suppose then we detotalize, decompose 'nature' to get a better understanding of it, and of how and why institutionalized understanding of it has enabled the emergence of contemporary climate change?

Philosopher Kate Soper's *What is Nature?* establishes a particularly valuable framework in this regard. She differentiates 'the "metaphysical", the "realist" and the "lay" (or "surface") ideas of nature', articulating these as follows:

1. Employed as a metaphysical concept ... 'nature' is the concept through which humanity thinks its difference and specificity ...

2. Employed as a realist concept, 'nature' refers to the structures, processes and causal powers that are constantly operative within the physical world ...

3. Employed as a 'lay' or 'surface' concept, ... 'nature' is used in reference to ordinarily observable features of the world: ... This is the nature of immediate experience and aesthetic appreciation; the nature we have destroyed and polluted and are asked to conserve and preserve.

(Soper 1995: 155–6)

Although Soper's first and third definitions will come into play in this book, to critique the way that poems such as Motion's elide the third definition with the first, it is the second, 'realist concept' of process that is the governing interpretation for my argument. This is because climate change reveals nature to be neither separable from us, as in Soper's first definition, nor consisting in what is 'observable', as it memorably becomes in Buell's initial emphasis on the role of mimesis in nature writing. In Soper's words, the 'realist concept' represents a nature 'indifferent to our choices, [that] will persist in the midst of environmental destruction, and will outlast the death of all planetary life' (ibid.: 159–60). This definition reveals that human and material phenomena are alike susceptible to the same principles; climate change itself should also put into doubt the idea that nature remains a category separate to the human, as in her first definition.

Nevertheless, Bate suggests that it has been a condition of our culture that we make just such a distinction, and draws on Latour to claim that 'The modern Constitution was above all premised on a strict separation between culture and nature' (Bate 1996: 439). Latour himself describes this 'constitution' as the 'common text that defines this understanding and this separation'; but on such a foundation *We Have Never Been Modern*, according to the title of his book (1993: 14). Latour argues that we have never fulfilled the condition of modernity that we imagine we have (a key theme in the work of T. S. Eliot, as I will discuss), because the presumed separation of nature and politics means we overlook the 'hybrids' generated between the two categories: 'All of culture and all of nature get churned up again every day',

Latour suggests (ibid.: 2). Much of Latour's work is concerned with this falla-
cious separation of science and politics, as distinct means of describing and
controlling nature and society respectively. He argues instead that the two are
interdependent and that their relationship creates 'quasi-objects' (ibid.: 51ff.),
entities that are neither essential nor constructed, neither objectively existing
nor subjectively perceived, but which share the qualities of both. Operating
according to the principles of Soper's second definition of nature and only
problematically manifesting according to her third, climate change breaks
down her first to generate Latour's hybrids.

In *Politics of Nature*, Latour maintains that 'People have been much
too quick to believe that it sufficed to recycle the old concepts of nature
and politics unchanged, in order to establish the rights and manners of a
political ecology' when instead these terms want 'a thoroughgoing rethinking'
(2004: 2): where politics (and indeed ecocriticism) behaves as though Nature
is fundamentally a source of truths, and society then uses its observation
or derivatives of these truths as the basis for its decisions, that basis is false,
Latour contends. He argues that 'Politics has to get back to work *without* the
transcendence of nature' (ibid.: 56; author's italics). We therefore need to seek
literary paradigms that do not depend on but challenge that transcendence,
and our place within it. Modernism, responding to an ever-increasing number
of scientific insights into the world – at the level of the principles Soper
describes in her second definition – and the application of those insights in
both industry and conflict, offers such a paradigm.

If ecocriticism is to attend to these concerns rather than dismiss Modernist
writing as insufficiently green, it needs to be more critical about its conception
of 'eco'; and it would do better to consider what Latour says 'we are left with'
after we disavow this transcendence, the 'multiple associations of humans and
nonhumans' (2004: 46). His remark provides the intellectual heft to enable
us to enact Buell's laudable critical requirement for texts to exhibit a 'sense
of [humans] being one among many actors in a much vaster and complexer
habitat', and 'to imagine nonhuman agents as [our] bona fide partners' (1995:
178, 179), even though his own attempt to do so chooses to neglect the agency
of CO_2 emissions. By restoring to nature the agency that Soper and Latour
enable us to think, we can interrogate the premise of much early ecocriticism
that assimilates environmental crisis to traditions of nature writing. While

conventional visions of Nature haunt us into the present – a thought that troubled Wallace Stevens, as I will show in Chapter 3 – the idea of tradition on which the recuperation and recovery paradigm of formative ecocriticism depends might become more radical if we consider not the history of nature or environmental writing but that of phenomenal agency and its entanglement with culture as constituting a neglected tradition.

To take the more particular example of climate change, we might consider the accumulation of greenhouse gases themselves a tradition; doing so would mean that climate change is usurping the role of the politicized literary scholar in bringing that history to our attention.[9] As the sociologist Ulrich Beck points out, 'the "side effects"' of such anthropogenic emissions, 'which were wilfully ignored or were unknowable at the moment of decision, assume the guise of environmental crises that transcend the limits of space and time' (2009: 19). Contemporary 'Global risks are [thus] the embodiment of the errors of the whole industrial era', he writes (ibid.: 100). To regard the side effects of industrial activity as a tradition in their own right involves a transfer of agency from acts of cultural intention to their unintended consequences. Modernism's conscious engagement with the newly explored territory of the unconscious again provides an opportunity to broach such a topic, and in Chapter 4, in particular, I will look at how the unconscious is projected on to the landscape in the work of Basil Bunting.

To consider this contrast between intentional and unintentional agency critically, compare the way that Bate requires 'the language itself ... to do ecological work' (2000: 200) with Mike Hulme's assertion that 'we need to reveal the creative, psychological, ethical and spiritual *work that climate change is doing for us*' (2009: 326; my italics). Bate focuses on the text, Hulme on the environmental phenomena and their presence in culture. By thinking of carbon emissions as a tradition, we can flip the tacit ecocritical assumption that it suffices to think of nature as culture's other, as in Soper's first definition, and instead entertain the idea of culture as nature's other. Nature then takes on the position of enforcing an environmental hegemony rather than being the 'other' to a cultural hegemony. Its processes environ and entangle our culture and, as the emergence of climate change reminds us, they 'will persist in the midst of environmental destruction, and will outlast the death of all planetary life' (Soper 1995: 159–60).

Matter of concern

The effort of ecocritics can be better directed, then, to reflecting on our entanglement with these processes – that is, to think from the environment rather than just fitting it into pre-thought cultural or political themes and categories. In order to articulate phenomena that criticism in general and ecocriticism in particular have not previously found amenable, the discipline has participated in a broader, so-called material turn in the humanities. The field of material ecocriticism has been attempting to reimagine the categories of cultural critique in accordance with this understanding of multiple agencies, and thus makes available resources to tackle some of the intellectual challenges faced when engaging climate change. This also offers a fresh way of approaching Modernism, in particular its environmental imagination, as recent critics have begun to recognize.

By acknowledging that 'The concrete sense' of such materialist criticism 'is also expressed in Latour's actor-network-theory', the critic Serpil Oppermann explicitly understands the congruencies between the sociologist's approach and the emergent critical practice (Iovino and Oppermann 2012: 466). The material turn aims to establish that humans are positioned in a field of various influences that cannot be controlled, and may only be managed with limited success. Stacy Alaimo and other materialist scholars take first-wave ecocritics' attempt to restore us to nature and give it some critical sophistication, with reference to Latour's mapping of actor networks and Beck's sociology of risk. Alaimo herself makes the valuable observation that 'agency is usually considered within the province of rational—and thus exclusively human—deliberation'; even though, 'Alternative conceptions', she suggests, would allow us to 'accentuate the lively, active, emergent, agential aspects of nature' (2010: 143). The recognition that material forces have agency in their own right is summarized by Serenella Iovino in terms not unlike those of Buell's; she attests that 'Humans share [a] horizon with countless other actors, whose agency—regardless of being endowed with degrees of intentionality—forms the fabric of events and causal chains' (Iovino and Oppermann 2012: 451).

Having proposed that climate change represents both intentional human and unintended natural agency, I posit that these 'causal chains' offer a productive way in which to describe its emergent phenomena. This understanding of agency has two consequences for Iovino: 'the first is that an ontological vision based on the superiority of human agency over the nonhuman "world of things" becomes problematic. The second is that we have to redraw the boundaries of the "self"' (ibid.: 457). We need to understand that, just because we haven't intended to change the climate, this does not mean that it is not still changing; this also means we can refute what Alaimo calls 'the astounding right-wing denial of global warming', which egregiously maintains that 'we have been granted the right to choose whether or not we "believe in" global warming, as if (quasi-religious) beliefs or personal opinions could insulate us from the emergent processes of material/political realities' (2010: 16). In contrast to such an individualist approach to climate change, the network of agencies actually involved in the phenomena is typical of the subjects considered by material ecocriticism in Oppermann's analysis. She writes of the 'multiple interacting systems, such as climate change[,] entailing geopolitical and economic practices' behind these 'unpredictable changes' (Iovino and Oppermann 2012: 461).

Material ecocriticism is not in itself the global warming criticism that Bate heralds; but in its breakdown of conventional categorization it does productive work for this project. It demonstrates that traditional literary categories are problematically placed to engage with the complexities of climate change, because we cannot consider the climate as a single phenomenon that is separate from us and can be mimetically represented in language. This occasions Alaimo's observation that

> we need to mark the limits of our own ability to render the material world with language. Such a sense of limits does not pose nature as exterior to human language, but instead acts to ensure an awareness that the process of making meaning is an ongoing one, a process that includes nonhuman nature as a participant rather than as an object of inquiry. (2010: 42)

Alaimo's remarks are especially useful because these are also the concerns that occupy Modernist thought and the poets I will be considering in subsequent chapters, in particular Wallace Stevens. If we are then to bring climate change criticism into being, we can still begin with Bate's appreciation that different climates of reading will affect our understanding of texts of the past. But rather than frame a particular canon of writing as being amenable to this approach, we should be able to take any given text as a nexus of the various forces described by material ecocriticism and examining how they interact, rather than demand that they comprehensively reflect reality. In highlighting that 'objects appear associable with one another and with social ties only *momentarily*' (Latour 2006: 80; author's italics), Latour reaffirms that it is impossible to achieve such a total effect, and highlights the role of the imagination in understanding as much as we are able to understand.

One of the methods of exposing agency he outlines is to use 'the resource of fiction[, which] can bring—through the use of counterfactual history, thought experiments, and "scientifiction"—the solid objects of today into the fluid states where their connections with humans may make sense' (ibid.: 82). This weighs in on the side of creative, literary engagement with climate change rather than mimetic, instrumental readings, in particular if we think of poetry as a self-conscious fiction, of the kind that Wallace Stevens, for instance, imagines. Latour himself maintains that 'sociologists have a lot to learn from artists' (ibid.), and it is to these Modernist artists that I now turn to see how they can help us address the problems I have outlined.

Modernism matters

The conditions of the early twentieth century marked seemingly unprecedented historical changes and a break with the past, which Modernist writers both forced and resisted to different degrees, and both that context and their response offer an opportunity to reconsider the tradition and function of literature in the time of climate change. The value of such reflection now is in no small part because these writers properly scrutinised what the Romantic inheritance meant as part of the literary tradition in an industrialized, globalized world, rather than recovering it in the hasty and unquestioning way in which early ecocritics appeared prepared to do. Free from typical or topical associations with nature, Modernism does not become a tradition to which to annex the writing of climate change, but an adaptable, developing mode that does not perpetuate nostalgic visions. Modernism does not forgo the chance to see nature's complexities and their extensive implications for the state and future of our planet.

My contention that Modernist-inspired modes of writing are better equipped to articulate climate change and its complexities has some, albeit limited, precedent. Richard Kerridge suggests some of the techniques that Modernism makes available to contemporary climate change literature, proposing 'that contemporary neo-Modernist writing has specific equipment for reaching into this subject, as writing that keeps to the personal voice and the conventionally poetic has not …[;] neo-Modernism … can bring into poetic space kinds of discourse not normally available to the personal lyric' (2007: 133). This touches on one of the key distinctions that I will be addressing: between formal engagement with perceived breakdowns in order in the modern experience, and a 'personal lyric' premised on stable conceptions of selfhood.

Poet and ecologist Mario Petrucci elaborates on how experimental literary form can express our relationship with the contemporary global environment, arguing that if our 'processes of perception and representation … are marred and distorted by being trammelled into certain stock ways of expressing oneself and understanding oneself,' then we run the risk of missing 'all the things one has to understand, know, experiment with (along with those we *can't* know, or at best merely glimpse) in order to be completely human, to be fully related to everything that happens to us' (Petrucci 2009; emphasis his). He continues: 'after Modernism (and is it really over yet?), we've got very considerable resources, templates and exemplars of how to work more fluidly with language, to reach the deeper truths of how it functions and expresses our relationship with ourselves, our relationship with creation and perception' (ibid.).

Even Buell argues that, when we are seeking such 'a thoroughgoing redefinition of the self in environmental terms',

> It might seem that modernism had made such a redefinition easy. For the[se] adjustments in persona, prosody, and image … have certainly to a large extent been enabled by such interdependent modernist cultural revolutions as the breakdown of trust in an autonomous self, the deterioration of faith in a symbolically significant universe, and a rejection of bound poetic forms. Under such circumstances, one might suppose that nothing would come easier to a late twentieth-century consciousness than imagining human selves as unstable constellations of matter occupying one among innumerable niches in an interactive biota. (1995: 167)

Yet Buell concludes that actually this 'is not the case', suggesting that 'It is [still] hard not to care more about individuals than about people, hard not to care more about people than about the natural environment' (ibid.), just as he turns away from consideration of CO_2. Buell is not wrong in identifying the difficulty of Modernist forms, but in disavowing an engagement with them and their possibilities he aligns himself with the return to realistic or lyric conceptions of the self in later twentieth-century writing, whose conventions lead to problems for the engagement with climate change in 'The Sorcerer's Mirror'. The resistance to non-realist aesthetics leads Buell and other ecocritics to assert particular versions of environmentality that are not able to accommodate or articulate the complexity of climate change. My objective therefore is to expand both on Kerridge's and Petrucci's observations and on the initial insights I made on Presley's poem, to offer an alternative, Modernist reading of environment in general, and climate change in particular. I will identify and analyse tendencies and techniques in Modernist literary aesthetics, specifically in the strains pioneered by Ezra Pound and by Wallace Stevens, and the early twentieth-century concerns in response to which these aesthetics emerged.

What enables this analysis is a recognition that the intellectual challenges with which climate change confronts us resemble the concerns of Modernism in both character and magnitude, albeit with considerable additional urgency. Perhaps the first articulation of the theoretical problems that Modernist texts pose for ecocritics also intimates some of the value that avant-garde writing can have for articulating the complexities of climate change. Carol H. Cantrell's '"The Locus of Compossibility": Virginia Woolf, Modernism, and Place' suggests that Modernism 'would seem to be hostile territory for a student of literature and the natural environment', because its aesthetics have 'taught us to privilege the formal and the abstract over the referential', while its exponents 'are famously expatriates, wanderers, exiles ... rather than [rooted] in local and national traditions' (2003 [1998]: 33). Even in this analysis of apparent opposition, we can see what are useful resources for engaging with climate change precisely because the phenomena of climate change exceed the 'referential', and require the kind of globally networked analysis we can glean from 'expatriates, wanderers, exiles', such as Ezra Pound, T. S. Eliot and Basil Bunting, rather than those only or largely defined by close or rooted relations to particular places as in the Romantic tradition. Modernism's very oppositional quality is valuable, as anthropogenic climate change is likewise resistant to received ideas of nature.

Cantrell argues that Modernists were reflecting on a changed understanding of nature in the early twentieth century and that 'it seemed not only possible but necessary to create or invent new ways of seeing, new ways of registering the perceptual shock of change, new ways of being readers and viewers, and to respond with a new urgency to questions about the consequences of human creativity' (ibid.: 34). She goes on to claim that 'Key elements of modernism—the attack on dualistic thinking, the foregrounding of backgrounds, the exploration of the relation of language to alterity, and the self-referential nature of symbol-making—are vital areas of inquiry for those of us who are interested in the relationship between literature and the natural environment' (ibid.). With these remarks, she significantly surveys aspects of Modernist aesthetics rather than subjects of representation that offer potential for ecocritical engagement, in contrast to the mimetic preoccupations of early ecocritics.

Cantrell also manages to situate her argument doubly in the Modernist moment and among its contemporary resonances by contrasting the rhetoric of intended meaning with unintended consequence. She maintains that

> Particularly in this [i.e. the twentieth] century, we have learned to enforce meaning and unity on large parts of the world by turning them into abstract spaces ... Yet even an extreme rationalist relationship with an environment is a relationship, though it is not seen as such, and it proceeds from and leads to further relationships, many of them unintended ... Modern[ist] writers saw that the disastrous world they came to inhabit was the result of choices made at very deep levels of creativity—including the level of perception—and their work gives us the chance to explore some of the unexamined ways in which we are making and unmaking the world at every moment. (Ibid.: 39–40)

Cantrell uses terms such as 'unintended' and 'unexamined', opposing them to 'enforce[d] meaning and unity', that recall Beck's sociology of risk. As material ecocriticism shows, the nonhuman processes of the world resist, with agency but without intentionality, the attempt at mastery also identified by Cantrell. Like Latour, Modernist writers also recognize the inherent contradictions of modernity, and present counter-narratives to those of progress and development.

Anne Raine picks up on Cantrell's points to state more explicitly that 'many of the texts most often considered modernist—those that seek self-consciously to disrupt established literary conventions—are valuable for ecocriticism precisely because their responses to modernity involve a productive questioning of conventional ideas about nature' (2014: 102–3). She relates these to more recent (eco)critical projects such as those I examined above. Although Modernist works

> may not speak very directly to the concerns of nature-endorsing ecocritics, these texts anticipate recent efforts by scholars such as William Cronon, Donna Haraway, Bruno Latour, Tim Ingold, Jane Bennett, Stacy Alaimo, and Timothy Morton to develop forms of ecological discourse that complicate, critique, historicize or abandon the concept of nature while taking serious account of the agency of nonhuman beings and phenomena. (Ibid.: 103)

More succinctly, in his book-length study of nature in the Modernist novel *Green Modernism*, Jeffrey Mathes McCarthy proposes that 'Now one can follow material criticism … to theorize about modernism and nature and better wrestle theories of the mind in relation to matter' (2015: loc. 21.2). He goes on to maintain that 'the benefit to environmental criticism is that modernism directs readers to the limits of mimesis, and lingering on these limits helps ecocriticism synthesize the confident representation of the physical world with the certainty that all perception is mediated' (ibid.: 72.6).

While both Raine and McCarthy are considering more generally the scope of ecocriticism and its potential for engaging with Modernism, the problems and possibilities they identify relate to my ongoing argument that modernity's shadow tradition of greenhouse gas emissions is implicated in the emergence of anthropogenic climate change. Modernist formal innovation often seeks to express or enact the contradictions and paradoxes at the root of this complexity, and thus its aesthetics remains vital in the present moment, in which new articulations of humanity's entanglement with the phenomenal are so badly needed.

In the next chapter, I will assess in greater detail the correspondence between these two contexts a century apart. Using environmental thinking and political ecology to read Modernist poetics and poetry, I will also use the poetry reflexively to read the critical discourse and inform the formulation of my methodology for climate change criticism. My readings of particular work in the Modernist tradition, beginning with T. S. Eliot's *The Waste Land* in the next chapter and moving on to consider Wallace Stevens, Basil Bunting and David Jones in those that follow, aim to develop a new understanding of both the poetry itself and my own critical apparatus. Finally, I will explore the climate change poetics that emerges in contemporary poetry, to which I turn in Chapter 6.

A New Climate for Modernism

Ezra Pound's exhortation to 'make it new' has long been seen as a guiding principle for the work of his close associates Eliot and Bunting, and Pound's influence can also be detected in the poetics of David Jones. But, as quoted in Pound's Canto LIII, the maxim itself also marks a response to the extremes of a changed climate. According to Richard Sieburth, Pound derived his imperative from the motto of Cheng Tang (Pound 2011 [1940]: 196, 325), coined during a seven years' drought to mark the emperor's attempt to redirect society through an explicit analogue with natural cycles, 'day by day' (ibid.: 196).

Schuster cautions that Pound's version of 'make it new', in contrast to Thoreau's use of the phrase in *Walden*, represents 'the epic poet matching stride for stride the epic ruler, linking art and state in producing a new culture that would also be a paradisiacal new nature' (2015: 154); he goes on to suggest that it 'implies a plenitude of seemingly inexhaustible resources' (ibid.: 155), both in Modernism's constant striving for innovation and in capitalism's drive for perpetual new product. As Schuster points out, though, in its original context

> the phrase was more a self-admonishment and humbling before the deity of the skies. The emperor reminds himself of the importance of water each time he washes his face with it. Waste and environmental distress are beyond the power of the emperor. Instead of the emperor having free reign [*sic*] to make things anew according to his will, or even to weed his empire so that its gardens persist, he must wait for the unpredictable rain to come in order to sustain imperial power. (Ibid.: 153–4)

In our current, changing climate, we should make 'make it new' new again with reference to this understanding of the environmental relations from which it originates.

Discussion of Pound's role 'in the emergence of a field of poetic practices that are ecological in both form and theme,' is attested by Raine, though she does not deny that 'The question of how to read Pound's ... generative influence on ecopoetics in light of the troubling politics of his vision of nature is a complex and compelling one' (2014: 102), not least because of his problematic relationship with the economics of his time and the implications of what Schuster calls his 'paradisiacal' vision. While not disavowing Pound's fascism and anti-semitism, I intend to scrutinize here the possibilities opened up by his poetics, which did influence poets on the Left such as Bunting.[1] Pound's conscription of the Chinese motto to his aesthetics suggests that high Modernist poetics might be bound up at their root with cycles perceived in the nonhuman world, 'an organic model of novelty' (North 2013) – as I shall show *The Waste Land* to be later on in this chapter. What remains important in Pound's vision is the simultaneous awareness of the nonhuman as both ever-new and perpetually so; it is part of a continuity at the level of Soper's second definition of nature, rather than a static, supposedly timeless tradition to which the writing of the environment might hark back.

Schuster contends that, 'We just don't know in advance what kind of ecological work will ensue from an artwork, but by studying the contexts and afterlives of modernist art, a sense of the conceptual and practical ecological stakes can come to the fore' (2015: 156–7). This recognizes, as Bate does, that changes in our environment, broadly conceived, affect our reading of texts of the past. When we read Modernism in this fashion, however, it responds in a very different way to Romanticism, or at least Bate's reading of Romanticism, because the possibility of being 'made new' is already under-stood by Modernism's exponents. The strategies by which they enacted this understanding make them distinctively valuable to the comprehension of climate change, and particularly valuable in their poetics are the following qualities: ironies of representation and a resistance to received ideas of 'Nature'; transnational or global scales; hybridization of natural change with cultural and social (anthropogenic) change and the breakdown of dualisms; a new problematics of environmental selfhood; language's vexed attempt to engage with the world and, reflexively, with its own materialism; and the expression of a troublesome environmental unconscious,[2] which has been repressed by narratives of civilized progress. I now turn to these strategies and expand on the relevance of each to ecocritical articulations of climate change, before going on to read Eliot's *The Waste Land* by way of demonstration of these principles.

The modes of Modernism

Irony: Changing climates of reading

Awareness of the shifting context that informs the text is part of the Modernist moment, as perhaps most significantly stated in T. S. Eliot's 1919 essay 'Tradition and the Individual Talent' (1975 [1919]: 37–44).[3] In the piece, Eliot proposes that the arrival of a new work of art affects our understanding of those of the past as it takes its place among them, implying that the reception of a literary work necessarily alters with time. In poetry itself, this principle could be taken to the extremes of repurposing works of the past to respond more explicitly to the present moment: again, Pound's *Cantos* offer a prominent exemplar of this practice, with their collation of texts from across history into an oblique commentary on the course of twentieth-century history. In composing the *Cantos*, Pound not only made the past new for the present, he also hit on a form that was itself open to expansion through time: as Michael Alexander points out, the poem's 'plans, even from the outset, included room for the topical, the new, the accidental; no end was clearly foreseen' (1998 [1979]: 128). Opening with the summoning of the dead by Homer's Odysseus (Pound 2011 [1925]: 127, lines 19ff.), the poet effectively allows each new voice of the *Cantos* to take its place in an ongoing sequence, enabling the entire work to span half a century and adapt to changing contexts. Following Pound's practice, both Eliot's collage of quotation and allusion in *The Waste Land* and Bunting's adaptation of the medieval Japanese work *Hōjōki* by Kamo no Chōmei offer more pointed case studies for the present argument, showing what it means to reconsider the literary tradition in the light of contemporary concerns and crises.

More generally, according to M. A. R. Habib, Eliot himself recognizes that 'The poet individuates by deploying the materiality of language, treating words as sharing the same individual material status as other objects in the world rather than as universal meanings or atemporal signs of objects. As such, a poetic construct will possess duration as well as unpredictability' (1999: 56). Thinking of language in this way, we can see that it remains open to different understandings through time by continuing to play its part in the networks of agency described by material ecocriticism, shaping and being shaped by changing conditions. This understanding means that, in using the term Nature, we must remember that it will never be a timeless sublime, a notion which is historically contingent on Transcendental or Romantic thought; nevertheless, as Buell remarks, many in the West still conceive of Nature in just such terms because of 'the inertial effect of the time lag between material conditions and cultural adjustment' (1995: 14).

In the work of the Modernists, the discrepancy between nature as conceived and as phenomena is made central rather than sidelined. For instance, Peter Nicholls argues that 'Nature ... is the prime deceiver' for the Modernists (2009: 22) because it not only fails to underwrite Romantic humanism or transcendent unity, but also cannot guarantee accounts of itself as ordered or beneficent. The resultant gap between language and phenomenon is where irony operates. Nicholls already claims that 'The quarrel with mimesis ... is often taken to define a pivotal moment of modernism's inception' (ibid.: 13). Enacting this slide between reference and referent, Modernist poetics is suited to expressing the irony that the climate in which we perceive and conceive of Nature does not correspond to those perceptions and conceptions.

Irony in this form makes language typical of human activity: while it purports to express one thing, it can effect another, and the potential for disjunction is greater according to the scale over which it operates. Paradoxically, the realization by a writer that the text is askew from the world at the point of composition informs that act of composition, and renders the text sensitive to and interrogative of the context – the reading climate – in which it is received. The destabilization of context through irony is what the text then communicates, rather than a definitive or timeless 'message'.

This can give poetry a crucial advantage over apparently more rational, technical discourses when it comes to engaging with climate change, in particular that of politics. The sociologist Bronislaw Szerszynski argues that political discourse remains mismatched with ecological crisis and the solution to this crisis is not to be found in a simple restoration of political language's reference to a reality outside language ... It was the cul-de-sac of modernity's 'correspondence' theory of truth – the idea that language and the world are separate, and that language can be judged by how it more or less corresponds to the world – that led to the crisis of representation in the first place. (2007: 338)

That is to say, politics attempts to engage with climate change by appealing, rhetorically, to reality; poetry, on the other hand, acknowledges that language is never able to achieve this end and instead explores its alternative resources for communicating the phenomena. The pronouncements of our leaders on climate change always betray a gap with these phenomena, but they have considerably less licence to explore that discrepancy – whether or not they would be willing or able to do anything about it if they could – than poetry does. Modernist poetics in particular grapples with this 'crisis of representation', which persists today in the gap between our individual experience of nature and the phenomena of climate change.

Globalism: Twentieth-century scale

Modernist writing articulates an awareness of these differentials in the ways that it offers access to scales larger than the human and the local. In *Modernism, Narrative and Humanism*, Paul Sheehan maintains, 'Scale prepares the ground for anthropocentrism' (2002: 6), and he goes on to assert that narrative 'is *human-shaped* ... to maintain the crucial human/inhuman distinction' (ibid.: 9; author's italics). This is an indictment of traditional narrative forms for their complicity in perpetuating the dualism between humanity and nature, sidelining the presence and agency of nonhuman forces with an anthropocentric world view that focuses on characters and, even at its broadest, on societies.

Sheehan continues: 'It is this very process, of course, that comes under increasing strain with modernist reworking' (ibid.: 13), remarking that the 'modernist novel liberates narrative's latent performative power by introducing formal irregularities ... Brokendown narrative is insidiously disquieting in ways that troubling story-content cannot match' (ibid.: 15–16). Because Modernism asks us to examine our modes as much as our subjects of representation, those modes can convey nonhuman agency in more productively 'disquieting' ways than if such agency were presented as the 'troubling story-content' of realist prose accounts.

Modernism seems almost constitutively to depend on reproducing and exacerbating the breakdown between human and global scales, doing what Cantrell identifies as 'jarring' readers 'out of routine habits of perception' (2003 [1998]: 37). It bears witness to an emerging sense of tension created between the scales of person and planet, and when we try to engage with those different scales in similar ways it exposes that tension. For instance, consider how two of the key questions of Eliot's 'The Love Song of J. Alfred Prufrock' (2015 [1917]: vol. I, 5–9) are both formulated in the same way: 'Do I dare / Disturb the universe?' and 'Do I dare to eat a peach?' (ibid.: 6, lines 45–6; 9, line 122). Eliot's poetics attends to the peculiar juxtapositions that modernity, in the form of globalization, brings about, through Prufrock's ontological uncertainty. The decision to eat a peach may also very well 'disturb the universe' when read today, if we factor in the water, soil and pesticides used to grow it and the carbon emissions associated with its shipment from the USA to the UK, for instance. This is characteristic of the difficulty of individual responses to climate change, where on a broader scale 'a person registers … less in terms of familiar social coordinates … than as a physical entity, representing so much consumption of resources and expenditure of waste (not the personality, but the "footprint")', in Timothy Clark's words (2012: 161). Hence we might now describe Prufrock's dietary dilemma in Clark's terms as 'absurd but intelligible' (2012: 151).

With this increasing uncertainty of scale, which brings previously remote influences into presence and makes connections between or across different levels look increasingly plausible, even the smallest of incidents risks deranging our sense of scale. In his essay 'The Noble Rider and the Sound of Words', Wallace Stevens writes: 'We are close together in every way. We lie in bed and listen to a broadcast from Cairo, and so on. There is no distance. We are intimate with people we have never seen and, unhappily, they are intimate with us' (1997 [1942]: 653). Stevens's 'unhappily' colours the rest of his remark with a tone of wary terror – the sanctum of the bedroom is penetrated by news from Egypt, and there can no longer be any secure privacy or peace.

Where distance becomes increasingly collapsed during the early twentieth century, our present inability to process persistent evidence and discussion of climate change bespeaks not only its remoteness as such but its ubiquity. John Lanchester writes:

> I don't think I can be the only person who finds in myself a strong degree of psychological resistance to the whole subject of climate change. I just don't want to think about it … Global warming is even harder to ignore [than the nuclear threat], not so much because it is increasingly omnipresent in the media but because the evidence for it is starting to be manifest in daily life. (2007: 3)[4]

What does it mean to offer 'psychological resistance', though, when climate change is pervasive and, of necessity, environing? The only way we can then look is within ourselves.

Identity: The unsustainable self

Rather than presume a stable personal identity premised on the 'familiar social coordinates' that Clark describes, in which individuals are able to alter their behaviour in the light of rational reflection on climate change,[5] Stacy Alaimo stacks up a range of conflicting forces to which the contemporary self is subjected: 'Humanism, capitalist individualism, transcendent religions, and utilitarian conceptions of nature have labored to deny the rather biophysical, yet also commonsensical[,] realization that we are permeable, emergent beings, reliant upon the others within and outside our porous borders' (2010: 156). Our paradigm must thus move on from the (pseudo-Romantic) individualism fostered by the concepts in the first half of her list, and show that Buell's abortive attempt at 'imagining human selves as unstable constellations of matter occupying one among innumerable niches in an interactive biota' (1995: 167) is not only worthwhile but essential to a poetics of climate change.

Modernism was grappling with a related identity crisis a century ago. Sanford Schwartz suggests that the recognition by nineteenth-century philosophy that 'there are as many "essences" as there are points of view through which to order experience' (1985: 18) is an important influence on Modernist aesthetics, with the effect that the figure of J. Alfred Prufrock 'may be so constructed that we apprehend the persona neither as a subject nor as an object but as a half-object' (ibid.: 197). That intellectual uncertainty was reinforced by the psychic trauma of the First World War in addition to the rapid technological developments that both enabled and were enabled by that conflict.

These uncertainties are further aggravated, if not rendered unresolvable, in the emergence of anthropogenic climate change; hence, the uncertain, 'half-object' status of personhood that Schwartz refers to in Modernism can be productively compared to Latour's conception of 'quasi-objects', hybrids of objective and subjective entities: 'Quasi-objects are much more social, much more fabricated, much more collective than the "hard" parts of nature, but they are in no way the arbitrary receptacles of a full-fledged society', he declares (Latour 1993: 55). Alaimo builds on Latour's analysis with an explicitly ecocritical agenda, recasting the idea of the 'quasi-object' as the idea of 'trans-corporeality, in which the human is always intermeshed with the more-than-human world' (2010: 2). She comments that, in relation to the effects of environmental hazards on the human body, 'trans-corporeal subjects must … relinquish mastery[,] as they find themselves inextricably part of the flux and flow of the world that others would presume to master' (ibid.: 17).

Modernist writers, in particular those who I am considering here, also sought strategies for negotiating between the poles of 'flux and flow' and 'mastery' that Alaimo identifies are characteristic of our present moment.[6] Scholars of Modernist selfhood, which is in search of its identity without the aid of a definite sense of traditional scale as Sheehan and Schwartz have noted, suggest that it is partly constituted by the threat perceived in external experience that they attempt to master. For Nicholls, the 'modernist aesthetic' of Eliot and Pound was designed to provide 'outlines and borders' with which the self can 'protect against the "chaos" of subjectivity' (2009: 192). Maud Ellmann uses terms similar to Nicholls, noting that in the work of the Modernist poets, 'the subject defines the limits of his body through the violent expulsion of its own excess' (1987: 94). By moving the discussion from the interior self to the physical body, she invokes the role of biological process in personal identity, implicitly locating the individual in their environment.

Alaimo in turn can provide the fuller environmental context of this process when she writes that, 'Forgetting that bodily waste must go somewhere allows us to imagine ourselves as rarefied rational beings distinct from nature's muck and muddle' (2010: 8); but Ellmann is already alive to the erosion of distinction between self and environment when she suggests 'The body and the city melt together' in Eliot's work, 'no longer themselves but not yet other' (1987: 99). This represents another continuity across, or transgression of, boundaries of scale, but also enacts Alaimo's 'trans-corporeality'. The environment is present through and beyond the semi-permeable membrane of the self. What borders exist around the Modernist self, then, do not necessarily connote an authoritative identity but actually point to the increasing pressures of the social environment that prompt the attempt to shore up selfhood, as in the case of Prufrock.

By revealing the tensions and contradictions in the self, the disjunctions of Modernism are a powerful way of grappling with issues fundamental to ecocritique, resisting the ease of both lyric polemic – 'we are victims' – and lament – 'we are perpetrators'. Prufrock could typify the figure of the wailing environmentalist, then, disempowered by his own knowledge in the way that Lanchester suggests he himself is. Prufrock is both seeing subject and seen object when he says 'I have known the eyes already, known them all — / The eyes that fix you in a formulated phrase,' and this double bind is the cause of his inaction: 'when I am formulated, sprawling on a pin, / When I am pinned and wriggling on the wall, / Then how should I begin …?' (Eliot 2015 [1917]: 7, lines 55–9),[7] a question that recurs in the circular motion of hesitation some ten lines later. Prufrock thus 'inhabits' what Schwartz calls 'the modern inferno where mere knowledge of one's condition does nothing to relieve it. The persona as perceiving subject is totally estranged from his own external actions' (1985: 201). In recognizing the self's implication in the world, Modernism is also reflecting a breakdown in a presumed wider order of that world.

Fragmentation: Hybridity and hesitation

With the transgression of traditional boundaries of scale, self and perspective, Modernist writers attempt to find ways to elucidate resultant connections and tensions, and their techniques can be correspondingly useful in establishing the relationship between human agency and climate change. Reflecting on the relationship between self and the world, Nicholls finds that Wallace 'Stevens regards the continuity between them as guaranteed by the imagination,' suggesting the poet's 'objective is to find in poetry some sort of equilibrium between these interacting pressures' (2009: 214). The exercising of imagination in this process is not purely abstract or fanciful, because it enables us as readers to make those intellectual leaps between the different scales necessary to properly comprehend the environment in crisis (as for example in the shifting timeframes of Presley's 'Triscombe stone'). Modernist poetry is configured to deal with just such relations, articulating them and the problematic context in which they occur through the use of brokendown style and unexpected juxtapositions. The result of this, Peter Howarth points out, is that 'the experience of actually reading a lot of modernist poetry is more like an immersion, where there is no longer a clear distance between what you are seeing and the position you are invited to see it from' (2012: 5) – an observation that demonstrates, by eschewing conventional forms of coherence, Modernist poetics is equally well equipped to enact the troubling phenomena of climate change.

The encroaching urban environment of modernity was after all a nexus or vortex of destabilizing forces. '[T]he rapidly expanding metropolis of the new era appeared increasingly unintelligible and contradictory', Nicholls argues, and in response, 'Writers could either retreat from it into pastoral fantasy … or they could plunge into the urban chaos' (2009: 16–17). It is in the recognition of the 'contradictory' forces that Modernism distinguishes itself. Nicholls regards 'an irony buried in the very frame of things' (ibid.: 22), and it is this recognition that comes to bear in the writing of climate change. This compares with Szerszynski's observation that irony has become a philosophical mode in modernity, where 'there are no separate groups of perpetrators and victims', and 'unlike conventional situational irony, there is no distanced observer, aloof from the folly and blindness they perceive being played out in front of them' (2007: 348), because we are all responsible, to a greater or lesser degree, for our contemporary climate.

Szerszynski writes that such irony 'finds expression in the very form of the modern novel, with its exploration of the multiple, incommensurable points of view that constitute any human situation' (ibid.: 340; author's italics). He proposes 'that environmentalist practice should acknowledge the debt it owes to aesthetic modernism, and more wholeheartedly align itself with that cultural current'; given that this would 'value and proliferate "impure" and vernacular mixings of nature and culture, new shared meanings and practices, new ways of dwelling with non-humans' (ibid.: 350–1), manifesting the quasi-objects to which Latour attests, we can extend Szerszynski's diagnosis beyond 'the modern novel' to think of it as descriptive of Modernist literature more generally. Eliot's *The Waste Land*, for example, mixes different kinds of discourse – *inter alia* literary, mythological, meteorological – and provides different perspectives on the relation between the self-consciously human sphere, and this technique is not unique to that poem. To adopt Alaimo's terms, his work can be seen to inhabit 'trans-corporeal space, in which the human body can never be disentangled from the material world, a world of biological creatures, ecosystems, and xenobiotic, humanly made substances' (2010: 115).

Modernist poetry is characterized by formal hesitations and discontinuities. Howarth remarks: 'Without syntax to restrict the fragments' meaning to their immediate context ... they can now connect to each other in multiple and unexpected ways' (2012: 6). This offers a technique for reading the peculiar connectivities that awareness of global climate change brings. Individual actions occur in an ironic, often indeterminable relation with the environment, 'implicating seemingly trivial or small actions with enormous stakes while intellectual boundaries and lines of demarcation fold in upon each other' in the words of Timothy Clark (2012: 152). Howarth's readings of Modernism emphasize that 'the banalities of ordinary material are given artistic charge by being poetically framed by structures in which no item or sound is ever subordinated into mere detail' (2012: 25); in the present context, the emergence of climate change offers one such 'charge' to our 'seemingly trivial or small actions', and Modernist modes present a means of attending to such detail.

Anthony Mellors's account of *Late Modernist Poetics* describes the attempt to make sense of this tension between juxtaposed elements and experiences by looking to the possibility of an organizing myth that might operate in Modernist verse:

> Myth always remains as the horizon of meaning, the point at which historical facts *should* cohere. But this will to coherence is at odds with poetry that gets its energy from the symbolic irresolution of violently contrasting elements ... Whatever its symbolic origins, private or public, its *objective* condition is to

remain fragmentary, unstable and unresolved; energy derives from the act of reading (the reader's desire to piece together fragments) as much as from the paratactic nature of the text itself – otherwise the poem would be nothing more than a message. (2005: 33; author's italics)

Mellors's analysis places us as readers in the position of trying to reconcile text with myth 'as the horizon of meaning'. In terms of climate change, we are in the habit of isolating our personal behaviour from its wider environmental effects, but in the relational context suggested by Mellors, we must ask whether those actions, with what Clark calls 'the cumulative impact of their insignificance' (2012: 150), debunk the possibility of an organizing myth, most predominantly that of Nature.

The indeterminacy of a text as patterned or chaotic, coherent or fragmented is aggravated by Modernist poetics, and the text's 'meaning' is only settled at each reading by the reader's capacity to impose, or have faith in an author's imposition of, unity. This informs McCarthy's ecocritical reading of Modernism, 'when texts like *The Cantos* or *The Waste Land* disallow closure to stand'; he elaborates that, although we can 'underestimate the extent to which ... Pound and Eliot did offer closure in the form of myth that displaced a discredited rationality, [it] is correct that the key momentum [of their work] is away from closure' (2015: loc. 321.6–322.9). In addition to emphasizing the context of reading rather than the sole authority of the poet as a determinant of meaning, such formulations also offer a productive way of thinking the relationship between the nature we experience in person, as per Soper's third definition, and our understanding of it. Is our experience testament to forces sublimely other to humanity, as in her initial, philosophical definition of nature, or is it the result of biophysical processes, to which all entities and systems on the planet are subject?

The very question points to the recalcitrance of the material, to which we would on the one hand ascribe transcendence and on the other contingency. Considering Modernists' response to this tension, Nicholls contends that 'the sense of the "fleeting" and "contingent" is perhaps the definitive mark of the early grasp of the modern' (2009: 6). However, his choice of terms, 'definitive mark' and 'grasp', suggests that the tension persists and is inherent in our attempt as readers to stabilize such contingency in texts. The way that Modernist poetics is able to enact this indeterminacy and remain contingent on our climate of reading will be seen in my analysis of *The Waste Land*, and in particular through consideration of whether or not the poem evidences order, disorder or some other quality of organization.

Time: The presence of the past

Modernist works remain contingent on a changing context by internalizing the possibility of their relationship with the future, as exemplified by the *Cantos*. In so doing, they also present a distinctive sense of time, which does not assume progress into modernity from a constantly receding past. For example, reading Eliot's 'Sweeney Among the Nightingales' (2015 [1920]: vol. I, 51–2), Frank Lentricchia argues that the poem exhibits an 'understanding [of] the present as an expression of the past, not so much diminished as it is luridly continuous, gross realist texture undergirded by mythic narrative. Allusion is the acknowledgement of the presence of the past; allusion says cultures are haunted' (1994: 261). The poem describes its titular figure as 'Apeneck Sweeney' before moving further away from humans' primate heritage by associating him with 'zebra' and 'giraffe' (Eliot 2015 [1920]: 51, lines 1, 3–4). A similar movement can be seen in the transition from 'The silent man in mocha brown' to 'The silent vertebrate in brown' in the fifth and sixth stanzas (ibid.: lines 17, 21), the latter rendering the figure doubly inarticulate by referring to him only in terms of his biological subphylum. The poem eschews the notion that human evolution is congruent with the progress of civilization: while 'hothouse grapes' are cultivated by human ingenuity where they would not grow naturally, they are still eaten with the 'murderous paws' of a woman likened to a beast (ibid.: lines 20, 24).

The different temporal modes of evolution and civilization superimposed by Eliot in the poem are comparable to the discrepant timescales invoked by climate change. While the development of life and transformations of climate have not necessarily occurred at the same rate or on the same magnitude over the course of the earth's existence, to think of either we cannot be content with reducing time to our individual experience. Human evolution requires us to think of ourselves as the product of millions of years of change, while anthropogenic climate change obliges us to entertain the idea that at least 200 years of industrial tradition, intensifying the activity of perhaps a dozen previous millennia, will have a cumulative and potentially very sudden effect within this or the next generation.[8] We are poorly equipped to think about such changes if we only consider the immediate or intended effects of our actions.

Eliot's hybridizing of evolutionary time with the present moment produces distortions conducive to a poetics of climate change, and his technique paradoxically makes 'Sweeney Among the Nightingales' a better reading of human subjection to natural processes than contemporary writing that explicitly engages with climate change as a crisis of the moment, such as 'The Sorcerer's Mirror' in its invocation of a three-millennia spell as if to suggest timelessness. A Modernist poetics of climate change offers the possibility of being more charged precisely for being unaware of, or oblique to, this context. Serenella Iovino points out that 'the new materialisms oppose ... *a vision of agency as connected with intentionality and therefore to human (or divine) intelligence*' (Iovino and Oppermann 2012: 453; my italics). In other words, an intention to write about 'the environment' cannot be the only criterion for environmental writing: that intention itself will have unintended effects or interpretations outside the topic, while writing that is not styled as 'environmental' can nonetheless engage with the questions of perception that environmental crisis raises.

Writing before anthropogenic climate change is recognized, the Modernists do not have the ideological baggage that comes with that term, and the insights they offer into human–environmental relations expose tendencies, processes and relations in a world that is warming up to be ours. By taking this approach, I am capitalizing on the way Cantrell identifies 'the chance to explore some of the unexamined ways in which we are making and unmaking the world at every moment' (2003 [1998]: 40) in Modernist work. The generative openness of Modernist aesthetics can articulate the uncertainty of a world in which human activity has divergent effects at the personal, cultural, socio-political, economic and global ecological scales.

The changing climate of *The Waste Land*

As I have intimated, T. S. Eliot's *The Waste Land* (2015 [1922]: vol. I, 53–77) can be read as juxtaposing the fragmented agencies already discussed, with the result that it unexpectedly places culture in a field of nonhuman agency. I propose, therefore, that it can serve as an exemplar of the kinds of associations we need to reveal if we are going to comprehend the complexities of anthropogenic climate change. As Latour remarks, in understanding the agencies in play 'it will ... be necessary to *represent* the associations of humans and nonhumans through an explicit procedure' (2004: 41; author's italics), and reading Eliot's poem in the twenty-first century offers one means of doing so. I do not mean to propose that this act supplants the scientific data, modelling and research to which Latour makes reference, but that reading stages and allows us to consider the phenomena in ways that science and politics cannot.

Indeed, Alaimo writes: 'If poetry and science are both "languages,"' then their 'struggle [is] to make the invisible visible, the unknown known, the material sensible' (2010: 53). In addressing such relations, *The Waste Land* pertains in the twenty-first century because it helps disclose the hybrid cultural and material agency entangled in the emergence of climate change by breaking down 'realistic' representation. For instance, the poem attends to both humanity and its emissions: world and waste are compounded in its very title, and Eliot's juxtaposition of the two discloses that they must necessarily co-exist. Crucially, his poem is unable to sustain the boundaries between culture and the others that culture creates, just as, in Latour's analysis, modernity cannot enforce the split it imposes between nature and society.

Read in this way, *The Waste Land* reveals the traces of relationships between humanity and its environment and complicates the impact that each has on the other. The collocation of 'waste' with 'land' also exemplifies Heather Sullivan's 'dirt theory', which identifies how 'Modernity's ... efforts to conceal "dirt" in its many forms have encouraged urban residents to believe that dirty nature is something far away and disconnected from themselves and their bodies' (2012: 526). We can thus recontextualize Eliot's offer to show us 'fear in a handful of dust' (2015 [1922]: vol. I, 55, line 30) in light of the horror provoked by bodily contact with dirt. That dirt can stand for both the wilderness and the dead, the spatially and temporally excluded and repressed.

Contemporary climate change itself manifests the history of industrial waste – carbon dioxide and other greenhouse gases – that resists its own categorization *as* waste to demonstrate its agency in the terrestrial environment. *The Waste Land* offers a literary analogue of this process, showing how problematic it is to try to account for everything according to (intentional) human agency, because what culture discards remains stubbornly present and influential in the poem. It is a demonstration *avant la lettre* of Ulrich Beck's description of 'more uncontrollable … global inter-relations in a world that is increasingly merging into a single planetary unit' (2009: 121). The poem helps us to negotiate this complexity because it exposes networks of agency that shape the world, and the tensions between them, within the compressed extent of the poem. It serves as an example of the 'very elaborate … artificial situations' that Latour says 'have to be devised to reveal [human] actions and performances' in their full context (2006: 79).

Latour contrasts the exposure of human agency with that of 'objects, [which] no matter how important, efficient, central, or necessary they may be, tend to recede into the background very fast' (ibid.: 79–80). He therefore sets out five criteria for an actor-network-theory that would account for objects' activity and agency (ibid.: 80–2). These criteria can be productively used to read *The Waste Land*, because they show that a poem that has been considered in terms of its cultural significance nevertheless contains the traces of nonhuman agency, to which it is exposed by the poet's formal innovation. The sociologist's methodological apparatus enables us to perform such a reading, overlapping with the Modernist preoccupations that I have outlined above – namely irony, scale, identity, fragmentation and time.

At different levels, *The Waste Land* demonstrates or reflects all five criteria for actor-network-theory. Specifically, the poem is (stylistically) innovative; it is remote from us in time, pre-dating the scientific identification and popular understanding of anthropogenic climate change by some decades, but remaining open to unintended consequences; it is presented as being in a state of breakdown, when compared with conventional narrative and verse structures; it interprets its moment of creation, each allusion being simultaneously a particular reading of that allusion; and it exhibits a kind of 'scientifiction' in its use of mythology. For Latour, these criteria are separate points of access to objects' agency rather than cumulative, but they each contribute to an interpretative network for the poem. Elemental earth, air, fire and water are predominant in these interpretations, with changes in the state of water being particularly telling. Eliot allows us to trace these elemental connections and agency by foregrounding them in the poem rather than letting them 'recede into the background' (Latour 2006: 80). Working towards a network of interpretations, I shall consider each of Latour's criteria in turn.

'Rendered ignorant by distance': The 'dull roots' of environmental collapse

Latour's criterion is that objects studied by archaeologists or ethnologists 'stop being taken for granted when they are approached by users rendered ignorant and clumsy by *distance*' (2006: 80; author's italics) and then shed light on the society that produced them. In this respect, *The Waste Land* as an 'object' is a take on the world of the early 1920s that, while distant in time, participates in industrial and cultural processes that will create today's climate, and engages with the intellectual challenges that those processes present. The poem cannot advance a proposition about anthropogenic global warming, topical 'climate change', as poets of the twenty-first century might, yet the process of reading the poem in relation to climate change can be endorsed by Latour's politics of nature: he maintains that 'quasi-objects ... can no longer be detached from the unexpected consequences they may trigger in the very long run, very far away, in an incommensurable world' (2004: 24). Processes of industrialization and urbanization were well advanced by the time *The Waste Land* was written and, in the sense that our contemporary climate represents the 'unexpected consequences' of these, we can identify some trace of civilization's ecological entanglement in Eliot's era. Precisely because it was written before the identification of anthropogenic climate change, *The Waste Land* can offer a distinctive cultural perspective on the phenomena, being party to the processes that have led to this moment without originating in our present understanding of them. We now read *The Waste Land* in a context of climate change; but what might it mean to read climate change in the context of *The Waste Land*?

Robert Pogue Harrison originally made the connection between Eliot and environmental crisis in *Forests: The Shadow of Civilization*, when he wrote: 'The wasteland grows within and without and with no essential distinction between them, so much so that we might now say that a poem like Eliot's *The Waste Land* is in some ways a harbinger of the greenhouse effect' (1993: 149). Harrison's phrasing conflates the 'greenhouse effect' – now more widely understood to be the mechanism by which the atmosphere has retained heat throughout its existence – with human-induced global warming. Nevertheless, his observation offers the opportunity to move away from consideration of Eliot's geological and meteorological imagery solely in terms of cultural or emotional sterility. Indeed, by problematizing our conceptions of a definite Nature, external to and separate from civilization, climate change necessarily demands that we refrain from closing down our readings, especially if doing so excludes their environmental implications.

In reflecting on the poem's critically neglected environmentality, it is worth considering even as an aside one possible meteorological influence on its composition. Eliot began the poem following 1921's 'summer of drought – no rain fell for six months', as Peter Ackroyd attests in his biography of Eliot (1985: 113). But the value of this detail lies in its suggestion of the environment that came to bear on composition, without being reducible *to* that environment. In the case of *The Waste Land*, Eliot may have taken a particular, contemporary instance of drought but he abstracts it into the mythologizing mechanism of his poem – as Pound does with the drought that inspires Cheng Tang's exhortation for renewal in Canto LIII. Eliot's personal experience thus enters a context where it is associated with the poem's geographically remote landscapes, in particular the wilderness of the final section of the poem. These associations mean that allusions to the context of composition are never just personal and lyric, but resonant across larger scales of time and space, in the way that climate change implicates individual experience and actions in much wider global phenomena.[9]

The poem itself is open to associative readings that explore this relationship in its juxtaposition of urban civilization with its environmental hinterland – even when the former imagines it is separate from the latter. This disjunction is expressed by Eliot's attention to what civilization attempts to disregard as externalities; so we see that 'The river sweats / Oil and tar' (2015 [1922]: vol. I, 65, lines 267–8), the side effects of the Thames' history of shipping. Perhaps more pertinently, the poem's being written in a time when pollution was more apparent in the environments of the urban West than it is now also figures the processes of industrial emission that a century of subsequent legislation – what Sullivan refers to as 'Modernity's many anti-dirt campaigns' (2012: 526) – have rendered less visible, if no less significant.

To register waste's presence in this way gives it 'not just a symbolic place but also [entails] a conscious and concrete embrace of dirt, which cannot be avoided since we live and breathe it daily' (ibid.: 517), making it an objective correlative for the invisible, insensible emission of greenhouse gases. My adoption of Eliot's own critical term 'objective correlative' here signifies, not quite 'a set of objects, a situation, a chain of events which shall be the formula of [a] *particular* emotion; such that when the external facts, which must terminate in sensory experience, are given, the emotion is immediately evoked' as he defines it (1975 [1919]: 48; author's italics), but a figure that points *away* from human interiority to everything that cannot 'terminate in sensory experience' but which still depends on our imaginative engagement to be understood: in this case, that is anthropogenic environmental change. The particular can express the general, the individual the environmental.

In the case of Eliot's queasy personification of the river, the physical presence of pollution – 'oil and tar' – indicates the flow of larger forces, both material and economic. Reflecting on the process by which these occur, Iovino remarks that 'Material ecocriticism considers the ocean as a porous body, a congealing of agencies and representations, of capital flows, life forms, "quasi[-]objects," and … of geopolitical forces, such as migration fluxes, or environmental phenomena, such as pollution and climate changes' (Iovino and Oppermann 2012: 457). In *The Waste Land*, the substances are 'congealing' in the 'porous body' of the river, thereby representing the 'capital flows' and 'geopolitical forces' that have brought them there. The poem's attention to the waste that human activity both generates and disregards also exemplifies philosopher David Wood's idea of a contemporary 'end of externality', where there can be no land without waste: 'externality is no longer available … Nature is becoming part of history in the sense that we are making irreversible impacts on the very processes that sustain its course' (2005: 173). Literary example thus precedes environmental philosophy; the disjunction that the poet identifies is part of the enabling history of anthropogenic climate change.

'Accidents, breakdowns, and strikes': Confusing the idea of order

Water is not only a medium of expression in *The Waste Land* because it also exhibits its own agency, running as an image through the poem and resisting any fixity that our perception attempts to impose on it. So for instance in the second part of the poem, 'A Game of Chess', Eliot juxtaposes 'The hot water at ten. / And if it rains, a closed car at four' (2015 [1922]: vol. I, 59, lines 135–6), making an immediate distinction between domestic utility and precipitation: while we utilize one kind of water for making tea or washing, we also cut ourselves off from the weather, and Eliot's juxtaposition is thus ironic. It reminds us that our lifeworlds form part of larger causal chains, in the same way that eating a peach may disturb the universe and a bunch of grapes can entail a hothouse. The 'closed car' demonstrates that it is civilized practice to keep ourselves dry in public, yet in the context of the entire poem, dryness is what will characterize the landscape of 'What the Thunder Said', where there 'is no water but only rock' (ibid.: 68, line 331), the culminating wasteland being ultimately the unintended result of the effort to minimize the effect of the weather on human society. By striving to keep itself dry, humanity is implicated in fostering the arid state that concludes the poem.

The poem's collation of different locations provides a context rich with potential associations and tensions such as this. In allowing hydrological imagery and terminology to run throughout, Eliot puts water-consuming civilization in closer relation with a parched, mythical landscape than everyday urban experience would allow us to recognize. Reciprocally, water itself cannot be kept free of human influence, as the 'oil and tar' sweated by the river indicate. The 'breakdowns' caused by civilization make obvious the water cycle that civilization disrupts, with discontinuities exposing the trace of connections, as Latour maintains; in *The Waste Land* they also draw us into the construction of meaning, engaging us with bigger, more abstract ideas at a scale beyond the human. The onus on us to follow the flows of water demon-strates that meaning in Modernist work can often be a function of readerly insight, however far authorial design directs that insight. So when Eliot's fragmented text prompts us to consider what order might be behind it, our faith in a Nature that could enable a unifying vision is today further reduced by our cognisance of the changing environment.[10]

'To bring them back to light by using archives': Dried specimens

The change in such an environmental reading through time can be comple-
mented by drawing on Latour's next criterion for actor-network-theory,
which is to consider objects in the light of their background 'using archives,
documents, memoirs, museum collections, etc.' in order to recreate 'the state
of crisis in which machines, devices, and implements were born' (2006: 81).
Taking a cue from these 'museum collections', we can reflect on the context of
Eliot's literary sources. The poem pointedly represents an anthropogenic state
of 'crisis' because its fragmentary quality is the result of deliberate decisions
by Eliot in response to comments on the poem from his wife Vivien and
from Pound. This understanding of an artificial crisis is doubly relevant in a
context where we have both unintentionally engendered climate change and
intentionally characterized its phenomena as a 'crisis'.[11]

To illustrate this creation of crisis from archival context, we can consider
Eliot's reference in his 'Notes' on the opening passage of 'A Game of Chess'
(2015 [1922]: vol. I, 73; note to line 77) to Enobarbus's description of
Cleopatra in her barge from *Antony and Cleopatra*.[12] However, in *The Waste
Land*'s version of the scene there is a confluence between Shakespeare's queen
and Petronius's Sybil, so that, crucially, the setting is an interior, removed from
the water – as with the 'closed car', Eliot's compositional process strives to keep
the text dry. One of the few words with liquid associations in Eliot's passage is
when 'synthetic perfumes ... *drowned* the sense in odours' (2015 [1922]: vol.
I, 58, lines 87–89; my italics), in contrast to Shakespeare's 'Purple the sails, and
so perfumed that / The winds were love-sick with them' (*AC* 2.2.203–4). The
artificial atmosphere in Eliot's verse overpowers sense in notably watery terms
– associating 'synthetic' substance with that water as 'oil' and 'tar' do in the
subsequent section of his poem – whereas the perfumed air breezes through
Shakespeare's lines.

Eliot thus creates a sterile modernity from an allusion to a fertile ancient
queen ('He ploughed her, and she cropped' [*AC* 2.2.238]) in a way that bears
out Robert Crawford's claim that this 'reworking of the Enobarbus speech
casts doubt on the validity of Shakespeare's interpretation of Cleopatra' (1987:
144). Rereading the poem today continues the cycle of interpretation by
offering a further layer on top of Eliot's,[13] and this practice of reading instances

the poet's own 'critique of notions of progress[, which] involved a perception that one myth comes to be interpreted by another, then by another again in a potentially infinite regress' (Crawford 2015: 183); hence, in *The Waste Land*, 'it seems impossible to find a point of view outside the process of history from which history can be interpreted' (ibid.). Crawford's sense of implication in history chimes with Szerszynski's remark that climate change makes it impossible to position 'the ironist as an outside observer of the irony on the moral high ground looking down, rather than implicated in it' (2007: 347). It emphasizes that in both the poem's landscape and on the planet, there is no point that can be isolated from any of the others.

This means that when other images in *The Waste Land*, such as the 'carvèd dolphin' (2015 [1922]: vol. I, 58, line 96), are like Cleopatra's barge abstracted from an aquatic context, we remain conscious of the water that runs throughout the poem and the effort taken to separate them from it. Eliot symbolically dehydrates his sources to connote their exhaustion, working in the theme of anthropogenic sterility that recurs throughout the poem, as for example with the abortion alluded to later in 'A Game of Chess': 'It's them pills I took, to bring it off' (ibid.: 60, line 159). Eliot channels his sources so that we perceive their artificial dryness when there was no such association in their original context.

Although similar concerns are in play in the 1923 collection *Spring and All* by William Carlos Williams, Eliot's US contemporary, comparing it to *The Waste Land* can show what is distinctive about Eliot's poetics in terms of their relation to natural phenomena. As John Felstiner remarks, Williams sought to respond to Eliot's 'feverish' vision, and express 'nature's renewable energy reaching deep into local ground with no stop in sight' (2009: 148). Rather than enact human failure to keep nonhuman phenomena at bay, Williams's poetics seek to enact those forces themselves, as well as their continual change: 'imagine the New World that rises to our windows from the sea on Mondays and on Saturdays—and on every other day of the week also' (Williams 2000 [1923]: vol. I, 178). His vision of nature is not timeless or inert, but renewed on a daily basis and, echoing Pound's credo, he ends his preface to the poems with the declaration 'THE WORLD IS NEW' (ibid.: 182). Again as in Pound, the change is cyclical, day by day; in contrast, doubt hovers over the dependability of any such sense of order in Eliot's poem. Moreover, Williams locates his poetry in space, 'the New World', rather than in the (old world's) passage of history, as Pound and Eliot do. The methods of the latter can be more valuable to climate change poetics because they are able to account for the haunting of the present by the practices of the past, which anthropogenic global warming demands we understand.

Rather than juxtapose the city with its hinterland and the present with the mythic past as Eliot does, the first poem of *Spring and All* swiftly abandons the few signs of human presence that it does contain: it is located 'By the road to the contagious hospital' (ibid.: 183), an image that suggests, as in Eliot's poem, the human attempt to contain and limit nonhuman agency. While *The Waste Land* cannot escape the human world even as material phenomena impinge on it, Williams, as though stepping from Eliot's 'closed car', enters the territory that the road sidelines, 'the / waste of broad, muddy fields' (ibid.). The past here is the immediate past of seasonal time, 'dried weeds' and 'dead, brown leaves', which foreshadow and participate in the renewal of spring. As such, while the emerging season is not Romanticized – 'sluggish / dazed spring approaches', 'One by one objects are defined' – it still carries a 'stark dignity', and none of the terror of the opening lines of *The Waste Land*. Indeed, rather than remaining shut out as it does in 'A Game of Chess', the wind in Williams's poem is 'cold' but 'familiar', amenable to human understanding. The first poem of *Spring and All* strips the natural of its mythical quality to identify the processes of Soper's second definition of nature, rejecting the Romantic impulse. At the same time, it denudes the world of human agency, whereas Eliot's own deployment of and reinterpretation of myth creates a fictive context in which we are able trace the human and the natural, the past and its consequences, as they collaborate uneasily and unintentionally in the creation of the world.

'The use of counterfactual history': Mythic time, clock time and climatic time

'The resource of fiction', writes Latour, 'can bring—through the use of counter-factual history, thought experiments, and "scientifiction"—the solid objects of today into the fluid states where their connections with humans may make sense' (2006: 82). While modernity aspires to arrest the dissolution of these objects, Eliot's poetics reshapes mythology into a distinctive fiction, one he plays off against urban civilization to show the deeper narratives in which those objects – and we ourselves – are implicated.

In the third part of the poem, 'The Fire Sermon', the proximity of the 'Sweet Thames' and the 'dull canal' (2015 [1922]: vol. I, 62, lines 176, 183–4, 189) juxtaposes a literary version of a river – a natural watercourse – with the grubby reality of an artificial one. But this contrast exceeds the material one of 'hot water' and 'closed car' in 'A Game of Chess', as the image of the canal-fisher also acquires a ritual association. 'Musing upon the king my brother's wreck' (ibid.: line 191) is not only an allusion to *The Tempest*, Act 1 Scene 2, but is also a reworking of pagan vegetation ceremonies, as Eliot indicates in the 'Notes' to the poem (ibid.: 72), so the line takes us out of urban time and into both cultural and mythic time. Eliot's contrast between modernity and mythology, in particular, can be read in the light of Latour's repudiation of clock time: the sociologist declares that political ecology 'has to modify the mechanism that generates the difference between the past and the future; it has to suspend the tick-tock that gave the temporality of the moderns its rhythm' (2004: 189). This is because the attempt to jettison the past and separate ourselves from nature marks the point at which hybrid phenomena, quasi-objects such as climate change, begin to proliferate, as Latour proposes in *We Have Never Been Modern*.

The institution of 'modernity' in contrast to nature creates an ironic disjunct that is then exacerbated in present-day seasonal shift. In *The Savage and the City*, Crawford is particularly attentive to the difference between mythic and modern time schemes in *The Waste Land*, asserting that 'In cities, where the seasons' impact is dulled, the rituals of fertility seem to lose their meaning, but they continue, processing like scenes in a play' (1987: 144), in a sham of their natural richness. The city actively downplays these cycles as modernity attempts to achieve historical progression, just as in 'The Burial of the Dead' a layer of 'forgetful snow' and 'sudden frost' initially prevents the growth of the past into the present (2015 [1922]: vol. I, 55, 57, lines 6, 73).

But the past cannot be comprehensively buried in *The Waste Land*, prompting the narrator's question to the figure of Stetson whom he encounters in the City: 'That corpse you planted last year in your garden / Has it begun to sprout? Will it bloom this year?' (ibid.: 57, lines 71–2). Eliot's likening of the dead to plants at this point anticipates Morton's observation that 'Nature is what keeps on coming back, an inert horrifying presence and a mechanical repetition' (2007: 201). For all that Williams recognizes these processes in *Spring and All* as well,[14] his mode of writing is one of quiet celebration rather than incipient horror; Eliot's poem instead acknowledges the repression that the civilized psyche enacts on nature. Morton argues that 'Environmentalism cannot mourn the loss of the environment, for that would be to accept its loss, even to kill it, if only symbolically. The task is not to bury the dead but to join them' (ibid.). That is to say, we must recognize our failed attempt to separate past from present, nature from culture, and accept the cycle of decay and renewal of which we are a part, rather than impose an ineffective division. In the figure of Tiresias, who has 'walked among the lowest of the dead' (Eliot 2015 [1922]: vol. I, 64, line 246), *The Waste Land* fulfils Morton's requirement and keeps us among the dead, to remind us of the shaping role of what civilization tries to exclude.

Eliot's poetics exposes such a project of demarcation as being fraught with ecological contradictions. His work exhibits the pervasive irony of Szerszynski's cultural Modernism, valuing and proliferating mixings of nature and culture (2007: 351) by using literary technique to challenge civilized practice and preconceptions. Civilization's attempt at isolating itself from nature leads the contemporary, commodifying mindset to separate 'hot water' from rain, domestic utility from environmental impact. In contrast to this, natural time jeopardizes an understanding of history as a progress away from environmental contingency, and is figured in the way the drowned sailor Phlebas 'passed the stages of his age and youth / Entering the whirlpool' (Eliot 2015 [1922]: vol. I, 67, lines 318–19). In the turning of the whirlpool, time is necessarily cyclical, cutting Phlebas off from the grounded narratives of anthropogenic time that civilization makes for itself, giving us access instead to mythic time. It seeps into the history of the poem despite civilization's attempts to suppress or ignore it, indicating our subjection to the deeper, material forces of nature.

In a critical book also entitled *The Waste Land*, Grover Smith uses the suggestive analogy of strata for the poem's organization of history, arguing that its landscape is 'made of interpenetrating layers of diverse cultural ages' (1983:

23). Once these are stratified, time should be able to take on its human significance as a series of new layers, landlocked and kept apart from the mythic time of the ocean. But no such separation is possible in the poem, given that water itself is mutable, appearing as snow, frost, rivers, seas, canals and clouds. Even at a moment when progress and civilized capitalism are signified by commuters making their way to jobs in the City, there is still a strong diluvian hint in Eliot's description 'A crowd flowed over London Bridge' (2015 [1922]: vol. I, 57, line 62); a bridge designed to cross a river is not up to much if it lets anything 'flow over' it. When the narrator then meets Stetson and enquires about the corpse, his final question is 'has the sudden frost disturbed its bed?' (ibid.: line 73). The cold snap has the potential to freeze the past, preventing its renewal into the present, while locking dangerous water into the form of ice and keeping the dead buried.

Winter surprises us by cropping up unseasonably often in *The Waste Land*. Habib comments that 'Eliot's presentation of the seasons as unnaturally disordered is marked with the history of human attempts to understand and control the seasonal cycle: the seasons themselves have comprised a universal point of identical reference, as indices of humanity's definitions of reality' (1999: 235),[15] as Buell has noted for instance (1995: 242). The disorder represented by contemporary climate change is testament both to the agency of wild nature and to its entanglement with our attempts to manage it, along with their unintended effects. The persistent winter of *The Waste Land* itself signifies the effort to hold water, along with the many processes in which it participates, in stasis.

Given this stasis, even water in certain states can become a form of separation or flood defence. Once we are conscious of this, it can refresh our reading of the opening lines of the poem. Spring, which is traditionally welcomed as the world's return to life, is here feared because of the 'stirring' caused by the rain it brings (2015 [1922]: vol. I, 55, line 3), the penetration of shoots from the ground, breaking through from the buried past: Eliot identifies civilization's repression of the material force that Williams in contrast celebrates in *Spring and All*. Reread in the twenty-first century, *The Waste Land* also offers a way of contrasting conceptions of seasonal order with the seasonal disruption for which climate change has been blamed.

'Innovations in the artisan's workshop': Exposing ecological process

The poem could have ended up colder still, however. To explain why it did not, I shall return to Latour's first approach for actor-network-theory: that is, 'to study *innovations* in the artisan's workshop' as 'one of the first privileged places where objects can be maintained longer as visible, distributed, accounted mediators' (2006: 80; author's italics). The published drafts of *The Waste Land* allow us access to the changes that Eliot made in response to Vivien's and Pound's comments on the manuscript and typescript pages. One of the processes thus traceable is that, by removing scene-setting passages, Pound ensures that the poem more jarringly juxtaposes its different locales, compacting its global scope and exacerbating tensions between its 'fragments', in particular between 'waste' and 'land'. A location afforded unusual prominence by the irregularity of the resultant structure is the seascape of part IV, 'Death by Water'; its brevity compared with the other parts of the poem paradoxically draws more attention to each of its lines. This was not the case during the poem's composition, however. Eliot's manuscript contains an account of a sea voyage that reaches the Arctic: 'And dead ahead we saw, where sky and sea should meet / A line, a white line, a long white line, / A wall, a barrier, towards which we drove' (2015: vol. I, 342, lines 549–51). This location, both a geographical and mythical end of the world, was excised by Pound in an act of editorial deglaciation. What remains in the poem as published are only the lines in which the body of 'Phlebas the Phoenician, a fortnight dead … rose and fell' in the currents (ibid.: 67, lines 312–16). With the disappearance of the icecap, 'Death by Water' becomes located between the Carthage at the end of 'The Fire Sermon' and the allusively situated desert of 'What the Thunder Said'. This jump-cut is a further instance of brokendown structure: it is not just the image of a drowned man that disturbs us about 'Death by Water', but the manner in which the poem presents him, adrift in mythic time.

Pound's removal of the Arctic context for *The Waste Land* increases the sea's level of importance as an image in the poem. Water floods the rest of the text, releasing the frozen past into its present, where ice had served as 'a wall, a barrier' in the draft. The water table that underlies the poem is likewise hinted at in Grover Smith's critical diction – he observes 'It is difficult to find lines simply stolen, not worked into the myth … because everything tends to flow together' (1983: 113). Reading water as the defining image of *The Waste Land* does not necessarily give it a structural role: its very changes of state and fluidity mean that it permeates the poem, rather than organizing it.

As there can be no clearly stratified progress from the buried past up into a surface-level present, our historical moment is as entangled in the processes that Eliot explores as his own time was, representing another breakdown in the boundaries civilization seeks to enforce. The hybrid, material agency of climate change thus imposes itself on retrospective readings. When we entertain climate change's implications to as full extent as we are able, they condition even our encounters with the canon; as for that matter other non-literary paradigm shifts have done, such as Darwin's theory of evolution. Texts are then reinterpreted in changing climates of reading so long as we carry on reading them. Advancing such an interpretative model, Timothy Clark remarks that 'The futural reading ... decenters human agency, underlining the fragility and contingency of effective boundaries between public and private, objects and persons, the "innocent" and "guilty," human history and natural history, the traumatic and the banal' (2012: 162). Such a decentring of human agency is dramatized through the course of *The Waste Land*.

Whether forecast

The tension between cultural and natural forces at work in *The Waste Land* raises the question of whether contemporary civilization is in a position to impose any successfully organizing principle on the nonhuman world. If there is a structure to the mythical nature that underlies the poem – in, say, the turn of the seasons – then there are grounds for saying that the 'damp gust / Bringing rain' (Eliot 2015 [1922]: vol. I, 70, lines 393–4) towards the end may offer the 'relief' longed for at the start of the poem (ibid.: 55, line 23). In the presumed course of the seasons, that relief could take the form of a wet spring following a barren winter, or a mild autumn after a summer heatwave. But already we have seen Crawford liken the urban seasonal cycle to a dumb show, while Marianne Thormählen asserts that 'the poem *begins* with the one explicit instance of regeneration it contains' (1978: 94; author's italics). Her implication is that a redemptive interpretation of *The Waste Land* cannot be sustained; in contrast, the first poem of *Spring and All* ends with hard-won regenerative promise.

The ambivalence of order makes *The Waste Land* a way for modernity to articulate and examine its environmental concerns. It can become an analogue for scientific uncertainty over to what extent we, in the second decade of the twenty-first century, are locked in to anthropogenic climate change by our historic greenhouse gas emissions and whether warming is taking place gradually or will happen abruptly.[16] In this context, the question of order is whether or not the earth's ecosystems will now sustain a state that can support human life, given the impact that humanity has already had on them. Crawford draws on the contested pattern of *The Waste Land* to offer two alternative interpretations of its conclusion, and the point at which they fork provides a staging of this environmental question: can the poem be read as offering us any chance of redemption, any hope that we will emerge from the final vision of the parched landscape? The first alternative that Crawford proposes is 'if the awaited rain fell at the poem's end, it would only lead back to that beginning, "breeding / Lilacs out of the dead land," with all its attendant suffering'. In this case, Crawford continues, 'The "Shantih" at the poem's end may be simply a way of stopping, [the] "formal ending"' to which Eliot refers in his notes (1987: 148–9; citing Eliot 2015 [1922]: vol. I, 77). The second interpretation is an 'exhausted collapse' (Crawford 1987: 149), condemning us to remain among the red rocks. The poem either rehearses the cycle of the seasons without sentimentality as Williams does, or it sees that cycle disintegrating.

I contend that we cannot make an assessment of what *The Waste Land* means today without drawing on our understanding of the future. The question is, then, whether we can attempt the radical transformation of our relationship with the environment, to maintain or restore contested ecological patterns, for all the 'attendant suffering' this will entail; or whether we will pursue our current course largely unaltered, until the point of our own exhaustion or a systemic collapse in the planet's ability to sustain us. To frame this question in more immediate, anthropocentric terms, has international climate policy built up enough momentum to ensure that it will achieve its admittedly modest ends? Or will the newfound recidivism of some states trump what belated progress has been made so far?

Where Harrison saw in Eliot's vision the recognition of the wasteland within, the environmental neglect of the twentieth and twenty-first centuries manifests the aridity of the poetic wasteland in the world around us. Or, to rephrase the critic's remark, what Eliot identified as a cultural malaise is now an environmental one as well, a pathetic fallacy made true; we are in the endgame of the 'Game of Chess'.

Wallace Stevens's Fictions of Our Climate

The poetry of Wallace Stevens makes frequent reference to the meteorological and seasonal, and the term 'climate' carries extra weight in his writing than its more straightforward, idiomatic usage. Stevens's persistent worrying away at notions of metaphor in his verse also elides easy distinctions between literal and figurative, reference and referent: Howarth remarks that the poet 'does not … allow you a secure vista on a scene from which you are safely excluded, and which then could not be believed in. Far more often, what starts as description turns out to be a series of shifting metaphors where the observer and the observed switch places' (2012: 132). Stevens's poetry expresses an awareness that we cannot separate ourselves from the world to regard it objectively, and recognizes that perception is always entangled with conception. Having already proposed that climate change demands a discourse that situates us within its complexity, multiplicity, contradiction and provisionality, I argue here that Stevens offers a poetics for exploring, managing and keeping in tension these different states of mind and world. Where the Eliot of *The Waste Land* breaks down civilized notions of order and locates their mythic roots, Stevens, tentatively, investigates how we might enable and further this re-engagement with the world through our imagination.

In this chapter, I will read Stevens's work as a reminder that the experience of nature cannot be separated from the human imagination. This recognition can be illuminated by Latour's notion of hybridity and the quasi-object to reveal a world neither entirely objective in its presence nor totally subjective as experience. Instead of assuming that we have easy access to an independent nature, Stevens continually returns to the question of how we constitute the world, acknowledging both the necessity and the pleasure of making fictive engagements with it. By failing to recognize this fictive, hybrid quality, civilization has supposed the reality of an externality, a separate 'natural' realm into which it can cast its waste, in particular the intangible emission of greenhouse gases. As a result, we make what is in Stevens's opus a metaphorical creation of nature into a material re-creation of it.

Stevens's poems enact the failure of language to master or contain the world, and deal both implicitly and explicitly with the climate's evasion of and resistance to our intentionality. He depicts the way humanity's imaginative intervention in the world aggravates non-intentional natural agency. In her ecocritical reading of Stevens, Gyorgyi Voros points out that 'If complete identification existed between human consciousness and "objective" world' – that is, if reference and referent were identical – 'there would be no need for language' (1997: 118), because perception would constitute immediate apprehension. She contends that we instead use metaphor because its 'efficiency lies in conveying a concept by way of an image' while 'its profligacy lies in that excess that spills over' (ibid.: 119), and it is the tension between the two forces that I will be exploring in this chapter.

I will first examine Stevens's use of seasonal and meteorological metaphor, and the implications of this for a poetics and criticism of climate change. His choice of imagery reads differently in the early twenty-first century because the terms in which we understand its vehicle – that is, climate – have changed so radically that they affect our understanding of his tenor. While the climate of Stevens's poetry is fictive, this does not make it false: its creative manifestations of abstract phenomena offer a cultural complement to the science of climate modelling, as I go on to explain. In concluding the chapter, I will use the terms of Stevens's own 'Notes Toward a Supreme Fiction' (1997 [1947]: 329–52) to suggest that the poems of our climate change 'Must Be Abstract', 'Must Change' and 'Must Give Pleasure'. I will read both this long poem and selections from Stevens's opus by considering these qualities.

Some poems of our climate

To focus on Stevens's use of terms such as 'climate', 'weather', 'season' and their associated lexicon might seem unproblematically descriptive or topical, if we were to take them as literal; likewise, we could be reductively symbolic if we consider them only in aesthetic terms. But Stevens's metaphorical meteorology makes the climate a product of both human and material phenomena, and indeed Scott Knickerbocker includes Stevens among those poets for whom 'Latour's hybrid notion ... in which nature is simultaneously real and constructed, better describes their conception of reality than a rigid dichotomy does' (2012: 9).

Stevens's metaphorical exploration of the human–climate relation anticipates climate change's material emergence as a product of that relation over the course of the twentieth century. To read the climate in this way fictionalizes it: not in the sense that the climate is a falsehood, but in that what are broader, illegible patterns and systems become more readily comprehensible in human terms. As in fiction, we see the significance of the general as it resonates in imagined particulars. The particular value of a climatic fiction is that it can acknowledges that our understanding of the world can only ever be provisional, and Tim Armstrong points out that for Stevens, 'Weather is a particularly engaging metaphor for that which resists order' (2005: 121). The climate then, considered either literally or metaphorically, is for Stevens the accumulation of conditions that prompt our imaginative engagement;[1] while we might usually refer to this accumulation as 'the world', that term implies a certain constancy and solidity, a dependable independence from individual agency, but this is hedged if we refer to 'the climate', already an average, abstracted from individual experiences. Stevens's work can be usefully read by material ecocriticism because both attend to the agency of our environments, which play as much of a part in constituting us as we do in constituting them, resisting our linguistic and cultural order. Where Eliot dramatizes this recognition through a collapse of received boundaries and collision of fragmenting discourses, Stevens enacts it in the reflexive examination of the imagination, in the attempt to distinguish what is characteristically human and what the human – as a self or as a civilization – projects on to the world.

Stevens's poem 'A Postcard from the Volcano' (1936)[2] offers one exploration of this relationship. The sun's presence in this piece symbolizes, for Helen Vendler, 'The persistence of nature' (1996 [1984]: 34), and Stevens contrasts it with the voice of someone who perished in an eruption. The sun is the largest source of energy for the terrestrial biosphere, and will not be affected by the deterioration of that system; as such, it bears out the persistence of one paradigm of nature, the causal but indifferent physical principles of Soper's second definition. But the narrator also informs us that the previous generation 'left what still is / The look of things, left what we felt' (1997 [1936]: 128). That is, manners of perception also persist, and along with the input of solar energy they come to bear on the way we view the world now.

Stevens problematizes our ignorance of the way these perceptions survive in language and their continued mediation of the solar reality, however: future generations 'Will speak our speech and never know' (ibid.: 129). The danger for us lies in our failure to recognize the hegemony that conventional ideas of Nature still exert, even when environmental conditions have radically altered; as Buell suggests, 'nature's prominence in the literature of the United States might be seen as only too conspicuous ... the inertial effect of the time lag between material conditions and cultural adjustment' (1995: 14). The 'dirty house in a gutted world' (1997 [1936]: 129) of Stevens's poem represents the ruined remnant of the Romantic sublime, and the sun's rays falling on it are not merely physical phenomena but golden, invested with human value, casting an ironic sheen of natural permanence over the historically ephemeral dwelling.

This entanglement of solar phenomena and perception, recognized by Stevens in metaphorical terms, becomes material with the emergence of anthropogenic climate change. His line break in 'left what still is / The look of things' invites us to linger on the possibility that a past generation has created a contemporary reality, 'what still is', until this is resolved as the appearance of a reality alone, 'The look of things' (ibid.: 128). Yet in our present climate, the subsequent line's 'spring clouds' (ibid.) troublingly represent the hybridized agency of humanity and atmosphere: how can we be certain the water vapour of which they are made comes from lakes and oceans, or from industrial emissions and aeroplane contrails? We have left things that *look* like clouds, whether or not they *are* the same kinds of cloud they once were. More than just the visual sense is required to appreciate our relation with, and implication in, the climate.

In the poem, the human enclosure of the 'dirty house' cannot shut out its cultural legacy, which takes the form of a 'spirit storming in blank walls'. The choice of 'storming' as a verb achieves a status somewhere between metaphor and description, thanks to the meteorological context Stevens has created in the poem. Read in the twenty-first century, where a 'tradition' of accumulated greenhouse gases in the atmosphere increasingly disrupts weather patterns, our own cultural legacy becomes manifest as a form of 'spirit storming' that likewise elides the categories of human conception and material phenomena and makes our dwelling on the planet less secure. In Knickerbocker's words, 'Stevens' conception of the imagination and reality as both distinct ... and inseparable provides a theoretical starting point for modernist ecopoetics, in which nature is simultaneously real and constructed' (2012: 24).

Seasoned selves

These transpositions between vehicle and tenor of metaphor also affect humanity's position in the world in Stevens's opus. His poems, Thomas C. Grey writes, are 'much more about places than about people. True, personae flock through them, but ... they do not emerge as living characters' (1991: 26). They are ghostly human figures open to the context of the imagined reality in which they are situated. While their status resembles Prufrock's semi-permeable selfhood (see Chapter 2, pp. 35–7), Stevens's figures are rendered with an evenness of tone that applies to all aspects of the world, equalizing human and nonhuman, rather than by the collision of discourses in which Prufrock is situated and by which he is constituted. This is evinced in 'The Snow Man', denotatively a 'body of the same substance as his environment' (Voros 1997: 139). In the poem's context, this Snow Man is far more abstract than the object of childhood play, however.

The poem's opening clauses, 'One must have a mind of winter ... And have been cold a long time ...' (Stevens 1997 [1923]: 8), might simply be describing (albeit with a hint of wry reflection) the conditions 'one' would have had to endure in order to witness this landscape, 'a glittering foreground—almost an obstacle—through which we are made to pass' (Vendler 1996 [1984]: 49). However, these observations are then qualified by what follows: one would have to achieve the conditions Stevens sets out yet still 'not to think / Of any misery in the sound of the wind', suggesting that these opening states are prerequisites for a perception that does not project human emotion on to 'the same wind / That is blowing in the same bare place // For the listener' (Stevens 1997 [1923]: 8).

Emphasizing the imperative quality of the initial 'must' and 'have been', we might read the poem as a challenge to survive in the wilderness, in which exposure to the winter and inhabiting 'the same bare place' is essential if the 'listener, who listens in the snow' is to nullify himself, achieving total immersion (ibid.) and the 'effectual abolition of that listener to a vanishing-point' (Vendler 1996 [1984]: 49). An interpretation such as this valorizes the wilderness over human identity. In contrast, we could read the 'must' and 'have been' as the projection of an impossible requirement, because the abolished listener would paradoxically have to retain the sense necessary to experience and articulate this attainment. The impossibility is hinted at by the repeated 'same' – 'the same wind / That is blowing in the same bare place' – because the need to assert that the man and land share weather and location indicates that we cannot intuitively experience that common identity other than through its naming; Stevens's repetition defamiliarizes the 'familiar' wind of *Spring and All*. Vendler identifies the discrepancy between sense and identity as having 'nature ... projected onto another plane, the plane of language' (1996 [1984]: 4). The Snow Man does then resemble Prufrock in that the environment of both comprises discourse, albeit in Stevens's poem a radically minimal form 'on the threshold between naming and abstracting' (Voros 1997: 50), compared with the conversations that Eliot's eponymous persona overhears. The environment of Stevens's poem is anthropogenic inasmuch as it is the aesthetic creation of the poet, and this complements the opening in which the cold creates the titular figure from his environment and generates the imagination itself, 'a mind of winter'.

However, Stevens does not in this poem necessarily endorse a pathetic fallacy in which wintry conditions stand for or evoke a passionless being;[3] weather, the other seasons and indeed all that does not admit of human control in his work share the intensity of the Snow Man's winter. In the second poem of 'Credences of Summer' (1997 [1947]: 322–6), for instance, it is not frigidity but its opposite extreme that is sought to disabuse us of the illusions we bring to our interactions with the world: 'Let's see the very thing and nothing else. / Let's see it with the hottest fire of sight' (ibid.: 322). The end-stopped lines have the spareness Stevens aspires to in our perception, yet the imperatives here share the ambivalent quality of those in 'The Snow Man': they are only necessary to urge us if 'seeing the very thing and nothing else' is otherwise impossible. The poem goes on to enjoin us to track the sun 'Without evasion by a single metaphor' (ibid.), yet it is only through the medium of metaphor that we can do so, by seeing it 'with the hottest fire of sight'. By trying to see the sun in its own terms, we have to adopt a solar intensity of vision, and we then *become* the environment in the way that the 'mind' of the Snow Man is 'of winter'. At the same time, we can only achieve that identity through the human device of metaphor: according to Knickerbocker, Stevens's poetry 'suggests that metaphor, rather than being merely ornamental, is fundamentally involved in one's perception of reality; in a seeming paradox, the closer we get to reality itself, the more metaphorical our experience of it becomes' (2012: 49). Given that they are 'Credences', these poems also show that even the supposed objectivity of seeing 'the very thing and nothing else' is a matter of belief rather than of superheated perception.

Stevens's paradox progresses in 'Credences of Summer' as the tone seems to shift from insistence to something approaching desperation, with the imperative to 'Fix' what is sought 'in an eternal foliage // And fill the foliage with arrested peace' (1997 [1947]: 323). The alliterative 'fix' and 'fill' stress the effort to render a natural truth static, while the metaphor of 'foliage' entangled in this consonance is far from being scorched by the 'hottest fire of sight'. The image makes an analogy between the human impulse to elaborate on truth and organic growth, so that in both imaginative and photosynthetic terms, solar energy drives change and foliation in the terrestrial environment. The contrasting desire to fix Nature as a timeless other favours a narrow vision of it that understands this fixity as a refusal to admit alteration. We are commanded to 'Exile desire / For what is not' (ibid.) not only because that desire exists, but also to make the present state, 'is', a permanent one, and therefore not to wish for a change of season. Although less intensely expressed, it suggests the terror that also opens *The Waste Land*. When we seek to restrain natural processes of change, we are at the limit of life and want to sustain ourselves in perpetuity.[4] Voros likens this state to 'a stagnant economy, in which all commerce between necessity and desire, barrenness and fecundity, summer and winter, sound and silence has halted ... there is no true fulfilment without both' (1997: 123). The tension between frigidity and fertility is the tension between anthropogenic order and phenomenal flux, which together produce the climate we inhabit.

The co-creation of world through natural phenomena and human intervention becomes more explicitly aesthetic in poem VIII. The sound of
a trumpet moves from behind the weather – it 'blows in the clouds' –
then 'through / The sky', before taking precedence as a herald, 'the visible
announced' (1997 [1947]: 325). It alludes to the New Testament's vision of
the apocalypse, when the 'the last trump' (1 Cor. 15.52) will signal the resurrection of the dead, making the hitherto invisible Kingdom of God present on
earth. But Stevens eschews specifically Christian imagery for a more abstract
formulation, in which the revelation is aesthetic rather than divine.[5] Music
becomes typical of art in this arrangement, because 'the visible announced',
as well as being the manifestation of weather patterns in the form of sky or
clouds, is, synaesthetically, the visual expressed as the auditory. Art, then, is
both the precondition of meteorological phenomena, ushering them into the
sky, and the condition of their reception or 'announcement' in language; it
is perception expressed, accessible to others and contributing to the way in
which culture receives nature.

But the trumpet arrogates a special status for its expression, because it first
supplements our experience of the seen world before proceeding to supplant
invisible phenomena. Later in poem VIII, Stevens writes 'The trumpet
supposes that / A mind exists, aware of division, aware / Of its cry as clarion,
its diction's way / As that of a personage in a multitude' (1997 [1947]: 325–6).
The remarks would invest us with the capacity to discriminate, by 'division',
between the music's position as 'clarion' or its manner of 'diction', and what it
heralds, were it not for the fact that this is supposition. By likening its manner
to 'a personage in a multitude', Stevens also gestures at individual identity as
a supposition, or fiction – and even one on the part of the instrument, as a
figure for art, rather than ourselves.

Compiling the climate

Stevens proposes that our perception of the world is always coloured by our
conception of it, and this recognition is doubly useful to an understanding
of climate change, because it enables us to identify that first, we still project
our fictions of Nature on to the nonhuman world, in spite of the fact that
– second – our physical and cultural entanglement in climate has exceeded

the possibility of distinguishing our artifice from processes such as the greenhouse effect or ocean circulation. In 'The Poems of Our Climate' (1997 [1942]: 178–9), Stevens scrutinizes the attempt to reduce the world to a clear image derived from nature, but in doing so exposes the tension between the categories of 'cultural' and 'natural'. The three pieces that comprise 'The Poems of Our Climate' express simultaneously an urge to bring the climate into focus and the necessity for the imagination to exceed that experience. Angus Fletcher considers Stevens as exemplifying the problems we face more generally when trying to render the scale and complexity of environments into text: 'If we turn to the poems of our climate as Wallace Stevens called them, we find extreme pressure put upon the classical aim of focusing image and action, and we ask how any reader could be expected to identify with the whole of an environment' (2004: 125). 'The Poems of Our Climate' enact our desire to reduce climate to something tangible, while acknowledging the failure of this process to capture the world in imagery and the subsequent need to keep reaching beyond it to comprehend the material phenomena that it comprises.

Stevens writes in the second of 'The Poems of Our Climate' that 'even … this complete simplicity' – as of the cold in 'The Snow Man' – means that 'Still one would want more, one would need more' (1997 [1942]: 179). Compared with the structural failure to contain the world under ice or snow in *The Waste Land*, Stevens lyrically recognizes the impossibility of uninflected blankness. The snow in the first of 'The Poems of Our Climate' has just fallen, but spring is approaching (ibid.: 178): we are in a transitional phase at the beginning of renewal. But the gradual emergence of spring from winter also emphasizes the starker clarification and reduction of nature in domestic environments, such as the indoor display of 'Pink and white carnations' (ibid.). The imagery concentrates the natural into interior particularity, with no verb to indicate the processes of which these flowers are a part, but then broadens out into the less tangible description of the indoor light 'more like a snowy air, / Reflecting snow' (ibid.). The introduction of the simile and the fussy need to recapitulate its comparison as 'Reflecting snow' disrupts the imagistic simplicity of the opening lines, suggesting the conceptual effort that is needed to sustain the autonomy of those images cannot itself be sustained. The movement shows that however absolute the image, it remains an incomplete account of the climate.[6]

Stevens's poetics contrasts with the concentration of material force in William Carlos Williams's 'The Pot of Flowers', which ostensibly describes a similar scenario; the titular flowers resist the light of perception as they 'take and spill the shaded flame / darting it back / into the lamp's horn', leaving the pot 'wholly dark' (2000 [1923]: 184). Reflection in 'The Pot of Flowers' is a reflection of objective agency rather than rhetorical strain: where Williams renders the recalcitrance of the object world into poetry to suggest that the sensible world is still not entirely comprehensible, Stevens traces, beyond the immediate presence of flowers, the processes of imaginative engagement that persist in trying to make sense of the phenomena that are out of the reach of immediate sensation.

It is difficult to reduce Stevens's flowers to their objective presence because they are already linked with human agency. As 'carnations' they are suggestively fleshy, but an association with human presence exceeds the linguistic and visual to become economic, because the flowers have been cultivated, picked and arranged to serve as a domesticated image of the natural. Moreover, these abstracted blooms must be unseasonably early given the snow outside, in which it is unlikely that they will have grown. Their presence therefore hints at greenhouse origins, or the transport and economic networks that bring them from warmer climes into the suburban milieu. What seems a simple signal of high spring, then, is managed by human agency to falsify or bely the lingering winter. Stevens was familiar enough with the practice of floristry to recognize that domestic blooms entangle both cultural and botanical agency: in 'The Bouquet' (1997 [1950]: 384–7), for instance, the eponymous flower arrangement 'stands in a jar, as metaphor', and Tony Sharpe cites this poem while observing 'that centuries of commercial breeding and hybridisation (which Stevens knew about) have made the rose naturally artificial, or artificially natural – for where does nature end and (horti)culture begin?' (2000: 68). The carnations of 'The Poems of Our Climate' are likewise not just images – they are images for imagery itself, the process by which our culture entangles itself with climate as soon as it tries to abstract (from) it.

In the third of 'The Poems of Our Climate', Stevens explicates the tendency that Vendler describes for 're-examin[ing] his premises anew in every poem' (1996 [1984]: 41): 'one would want to escape, come back' to the previously conceived image or thought. This procedure of 'escaping' and 'coming back', essential to his poetics, also has points of comparison with the contemporary practice of climate modelling. Simulations of climatic states have to be individually run, with each dependent on a defined set of input parameters, and are performed iteratively to compile a range of likely and less likely scenarios. Kendal McGuffie and Ann Henderson-Sellers suggest that these simulations are instrumental in that they are tailored to individual purposes: 'different model types are better suited to answer different types of questions' (2005: 241). But inasmuch as models are explicitly run in response to particular questions, they can also be anticipatory, determinative; Szerszynski argues that climate models prefigure certain types of technical or technological response, and are thus linear and instrumental: there is an 'always-already presumption of application' (2010: 19). Rather than working towards a specific output or scenario, however, Stevens is in 'The Poems of Our Climate' tending in the opposite direction, to investigate the imaginative impulse at the root of both scientific and poetic practice.[7]

To elaborate on this difference: while models are designed to generate particular answers, Stevens reminds us that 'The imperfect is our paradise', that there never has been a stable, Edenic state, and we shouldn't direct our efforts towards identifying or achieving one (1997 [1942]: 179). If we consider this in aesthetic terms, we could read Stevens's assertion that 'Since the imperfect is so hot in us', any pleasure we take 'Lies in flawed words and stubborn sounds' (ibid.). The ambiguity of 'lies', though, suggests that in taking such delight, we countenance a false account of the climate. This might be distinguished from a fictional account in that the false one would be judged against the possibility that it could or should have been literally true, whereas the fiction could help us to understand the phenomena while eschewing the possibility of absolute truth.

Furthermore, the line 'Since the imperfect is so hot in us' enables a reading of our mismatch with the climate in terms of thermodynamics: the imperfection of our linguistic and imaginative systems leads to metaphorical build-up of waste heat. But by imagining these cultural framings as identical with the world, our practices as a civilization build on that false premise to transfer entropy from metaphor into materiality, accumulating in the discontinuity between concept and phenomena. Our present, anthropogenic climate recontextualizes Stevens's thermodynamic metaphor by converting the vehicle into the tenor, because the waste heat and greenhouse gases of industrial process demonstrate the imperfection of artificial systems founded on misconceptions. 'The Poems of Our Climate' do not need to give this imperfection a moral spin to pinpoint the fallacy that human imagination is sufficient to the world.

Models for atmospheric apprentices

Stevens's poems reveal both the impossibility of enclosing the climate as a way of understanding it, and the continual necessity for us to do so to in order to bring it within the scope of our imagination. In his reading of 'The Novel as Climate Model', Jesse Oak Taylor proposes that (in the case of Dickens's *Bleak House*) 'rather than making the text irrelevant to the world, [the novel's] closure is precisely what renders the text legible as a model rather than simply a representation' (2013: 5). By contrast, reading Stevens's poetry as climate model can offer both a model of the climate, fulfilling Taylor's criterion for literary form as 'a performative model of climatic phenomenology—the experience of climate *as* climate' (ibid.: 4; author's italics), and a model of the process of climate modelling itself. Unlike the atmosphere that Taylor describes in Dickens's novel, the 'climate' of 'The Poems of Our Climate' is a provisional juxtaposition of images, for which its floral arrangement is synecdochal. In the brief extent of the lyric space, Stevens uses concision and repetition rather than novelistic detail to create his climate models; nevertheless, like Dickens, he 'perform[s] the experience of the climatic encounter, which is never purely material but is instead bound up with the linguistic and cultural construction inherent in the imaginative projection required to inhabit an abstraction' (Taylor 2013: 3–4).

The salience of Stevens's observation in a contemporary context is that scientific understanding of 'climate' is similarly an aggregate of variables as a working model. Szerszynski writes of these experimental processes that 'it is because the unruly, surd complexity of the weather is being tamed by being forced to pass through standardized forms of measurement, and through conventional practices of aggregation and modelling, that we are able to conceive of such abstractions as average global temperature or rainfall, let alone see them rising or falling' (2010: 22). Szerszynski's comments emphasize the instrumental direction of these processes: their function is to communicate abstractions. Poetry offers a complementary mode of communication because, while it also seeks to render the abstract comprehensible, it never purports to be more than indicative in so doing. The fictive quality of a poem expresses climate more readily than climatic phenomena do themselves because the text is oriented towards the human reader, while the evident artifice can also caution us from taking it to be a like-for-like representation. Taylor contends that 'Ecocriticism must embrace the power of mediating constructions, instruments, and models *as models* if it is to engage productively with the imaginative challenges posed by anthropogenic climate change' (2013: 2; author's italics). If we do not bring a like awareness to scientific discourses of climate, though, we – and in particular those of us without scientific training – risk making these themselves the focus of our attention rather than the physical phenomena.

For Daniel B. Botkin, when we commit this fallacy, 'huge climate models are [then] the theory itself, and there is little evidence, and some contradictory evidence, that this is a helpful approach' (2012: 339). He argues that, in contrast, what 'computer models can tell us is the implications of what we know (the facts) and what we assume about a system that interests us ... This is the best use' (ibid.: 277). By drawing out the factors we consider to be of relevance, the model allows us to examine our own assumptions. 'The Poems of Our Climate' attends to the implications of what we can know, as well as what we cannot, the elusive persistence of the 'unruly, surd complexity' of all climatic phenomena that Szerszynski describes. This contrasts with the practices of 'standardized measurement', which on his account seek to 'force' and 'tame' the world rather than recognize its resistance to our control. Botkin further argues that 'the harder we work to force environmental constancy onto our surroundings, the more fragile that constancy becomes and the greater the effort and energy it takes' (2012: 290).

For me to propose that the climate is a construction, then, is not to claim – as a denialist might – that it is merely a politically motivated supposition rather than material phenomena; it is to see that the attempt to reduce our understanding of climate to data is effortful precisely because it can never be defined or understood by those data alone. Without the processes that Szerszynski describes we would not be aware of the changes in physical climate, and their importance cannot be overstated. Taylor even points out that 'Rather than distancing us from the realities of climate change, mediation and modeling provide our only evidence of its existence' (2013: 2) – just as metaphor can bring us closer to the phenomenal. Stevens's poem recognizes that we always need to reach beyond language, even if we can only do so *in* language, to appreciate what outruns our understanding and management. Literary form is able to expose the resistant agency of the climate itself, because as Taylor puts it, in Latourian terms: 'That which was once "purely" an object becomes, in its impurity, a kind of subject: an active constituent' (2013: 13).

Climate modelling responds to the resultant uncertainty by generating new models, and this recursive quality is a function of each previous iteration's discrepancies. More sophisticated, integrated climate models do not limit themselves to considering meteorological, oceanic and other phenomenal criteria either, but 'explicitly (albeit qualitatively) incorporate economic considerations, [and] estimate anthropogenic emissions requirements' (McGuffie and Henderson-Sellers 2005: 243), directing the process towards ever-greater accuracy by the inclusion of ever-greater numbers of parameters. Botkin posits this as part of

> a kind of ecological uncertainty principle: The more you try to explain all the details, the more likely you are to make quantitative errors that lead you astray. The more details you seek to include, the greater the chance of errors that lead you astray. Yet, if you make your model (your theory) too simple, you are likely to miss the very qualities that determine what actually happens. (2012: 281)[8]

In contrast, Stevens's poetics of recapitulation responds to the uncertainties he finds by enacting the world's perpetual resistance to human mastery, from which the problem identified by Botkin stems. By accepting the contingency with which Stevens works, we may acknowledge that the climate is changing without that acceptance having to depend on definitive depictions of our future. A belief in the possibility of accurate modelling would also be a belief that we can fully comprehend physical processes and outcomes; which, even before we aggravated and intensified them, were already impossible to describe in their entirety, let alone manage.

Where poetry's sophistication differs from science's is that it frames our understanding of the world as provisional rather than progressive. To offer another example from Stevens's opus, 'Sea Surface Full of Clouds' (1997 [1931]: 82–5) contains five iterations of the eponymous scenario during an autumn cruise off the south-west coast of Mexico, eschewing the notion of absolute mimetic truth in each alternative vision to exemplify a bravura way of thinking and rethinking the world that is never satisfied with a definitive account. For instance, the ocean is, among five parallel phrases, 'the perplexed machine' and 'the dry machine' (1997 [1931]: 82–4). The term 'machine' is repeated mechanistically, making the Romantic sublime of the ocean into something more resonant in the industrial–scientific age. A metaphorical generator and a generator of metaphor, this 'machine' demands our fresh understanding each morning: Vendler suggests that 'The daily impersonal newness of the visible world was at first a disturbing thought to Stevens, as we know from … *Sea-Surface Full of Clouds*' (1996 [1984]: 59; author's italics). The poem also slyly rebukes globally northern conceptions of nature with its Pacific location, 'commemorat[ing] the illicit achievement of summer in November' (Sharpe 2000: 112).

The ocean's agency and potency shifts through the poem. It can be incipiently menacing, 'in sinister flatness' (Stevens 1997 [1931]: 83); it is likened to music, 'as a prelude holds and holds', aesthetically controlled and ordered (ibid.: 84); or it can even be just 'perfected in indolence' (ibid.: 85). Stevens's repeated recharacterizations of the ocean represent a response to the sea's own creation of itself, tracing the give and take of imaginative and oceanic agency.[9] The title of the poem is itself suspended between human perception and meteorological phenomena. The sea is 'full' of clouds from the point of view of an observer standing on deck, for whom their reflections appear in its surface. But the sea surface also comprises water that has come from clouds and that will evaporate to form new clouds, so in that material respect it is also 'full' of them. Both readings transgress the linguistic separation of 'sea' and 'clouds', according to our understanding of 'full' in terms of either perception or the water cycle.[10]

Just as its title resists easy determination, so the poem rejects definitive conclusion. Instead it draws attention to the changes in the world that, in a way paradigmatic of Stevens's work, requires we constantly re-examine our premises: from the combination of sea and sky 'Came fresh transfigurings of freshest blue' (ibid.), a corollary of Pound's drive to 'make it new' in the manner of the sun. As in *The Waste Land*, the effort to separate different forms of water – sea and clouds – proves futile because it transgresses a boundary and as such is a hybrid phenomenon. The 'freshest blue' in which this commonality of ocean and sky results signifies the renewal of imaginative energy and perception, aligning these with the natural cycles of waves and of night and day that have stimulated the poet's engagement. These cycles persist at the level of physical principle, and demonstrate that we have always been entangled in the act of creating and recreating the world; today, however, this is taking place in a significantly more material way than Stevens envisaged.

Notes towards a climatic poetics

To foreground the fictive quality of our engagements with the world is for Stevens a response to the way unacknowledged preconceptions shape that world: 'we live in an intricacy of new and local mythologies, political, economic, poetic, which are asserted with an ever-enlarging incoherence', he maintains in the essay 'The Noble Rider and the Sound of Words' (1997 [1942]: 652). Thinking instead of 'a fiction, which you know to be a fiction', as Stevens does in the 'Adagia' (1997: 903), is to acknowledge our impulse for ordering and understanding the world, while ensuring we do not lose sight of its potential for discrepancies, or 'incoherence'. I have outlined some of the myths of measurement and mastery that perplex our contemporary engagement with climate change, and would contend, with Voros, that Stevens's poems 'are the imaginative enactment of stepping outside … conceptions in order to create "a nature"'(Voros 1997: 35; citing the adage 'The poem is a nature created by the poet', Stevens 1997: 905). That is, we can create a model in which we can engage with our assumptions about the world. In their self-recognition and the relentless questioning and reframing of themselves, such fictions as a concept are valuable to our understanding of contemporary climate change.

By thinking of a 'climatic fiction', I do not intend to dispute the reality or the severity of the phenomena, neither human responsibility for them. The reverse is the case: 'climate change' is a fiction inasmuch as our use of the term, and our particular cultural framings of it, are insufficient to and exceeded by the material phenomena and their agency. Incorporating this awareness in our discourse of climate change better prepares us to deal with the inherent uncertainties of the phenomena than does an insistence on comprehensively verifying them. A 'climatic fiction' therefore acknowledges that its truth is not literal but is still necessary. It is a mode in which we can entertain manifestations of climate change as indicative of its presence, and develop the imaginative faculties that enable fuller engagement with its phenomena. This corresponds with Bonnie Costello's suggestion that in Stevens, the imagination can 'reveal the entanglement of nature and culture; the interplay between our desires, our concepts, and our perceptions; and possibilities for renewal and vitality within that entanglement' (1998: 574); a process that intersects with, but is not limited to, the practices of scientific and literary climate modelling.

Using the framework of Stevens's long poem 'Notes Toward a Supreme Fiction' (1997 [1947]: 329–52), I want to assess the value of a fictive climate and the possibility of deriving a contemporary poetics of climate change from it. Stevens's three criteria for his 'Supreme Fiction' are: 'It Must Be Abstract', 'It Must Change' and 'It Must Give Pleasure'. By 'abstraction', I want to consider the way poetry stands apart from the reader's immediate experience and challenges the priority of the personal in our conceptions, as our individual experience is at odds with our cumulative environmental impact. By the need for 'change', I maintain that poetry's invitation to be reread in a changing world can provide an adaptive quality for a climate change poetics. In the context of the 'pleasure' that Stevens requires of the supreme fiction, I will suggest, analogously, that poetry can make nature, at Soper's level of scientific principle, sensible to the reader, in the dual sense that the supreme fiction appeals to the senses and that it makes some, albeit contingent, sense of the phenomena.

Reality check: 'It Must Be Abstract'

To consider climate change requires that we engage with the climate as an abstract, beyond experienced weather. This is one of the crucial reasons *Why We Disagree About Climate Change*, according to Mike Hulme: 'Climate cannot be experienced directly through our senses ... climate is a constructed idea that takes these sensory encounters and builds them into something more abstract' (2009: 3–4); this construction occurs, for example, through the methods Szerszynski describes (2010: 22).

Stevens's poetry takes place between the particular of experience and the imagined abstract, mediating between the two. It allows the particular its weight while also giving substance to what is abstract, and his metaphor traverses the differing scales by which we need to understand our own and our culture's implication in the climate. Timothy Clark points out that 'The self-evident coherence of immediate experience, far from being the possible foundation of secure theorizing, is merely epiphenomenal and unable to see itself as such. It projects an illusory ground, a surface realm of human possibility, one that is delusory and even sometimes a form of denial' (2013: 12). Stevens's poems recognize this 'illusory ground' and offer strategies for negotiating it.

At two different extremes on the scale of terrestrial influence, the solar and the self are shown to be mutually creative agents in the first section of 'Notes Toward a Supreme Fiction'. The opening poem juxtaposes 'this invention, this invented world' with 'The inconceivable idea of the sun' (Stevens 1997 [1947]: 329) to suggest a contrast between a knowingly fictive environment and a sun beyond human conception. Indeed, the subsequent stanza exhorts an apostrophized youth to 'see it [the sun] clearly in the idea of it' (ibid.). But in inviting us to 'see it clearly in the idea of it', Stevens expresses a paradoxical solar absolutism, similar to that of 'Credences of Summer', since the line includes the anthropocentric 'idea' in its formulation. That 'idea' is more explicitly anthropogenic in this poem than in 'Credences', because, its clarity a result of cleansing rather than burning, it is approached through a metaphor that does not derive from the image of the sun itself: 'How clean the sun when seen in its idea' (ibid.). This co-constitution of world from solar phenomena and human imagination is confirmed by the entangled insistence that our star 'Must bear no name, gold flourisher, but be' (ibid.: 330), because 'gold flourisher' is bestowed as a name just as the poem has outlawed naming. Language, rather than the sun, is then the flourisher on being.

This doubleness of perspective, which yearns for objectivity but has no terms other than those of human experience in which to express it, recurs throughout 'It Must Be Abstract'. In the sixth poem, Stevens defines the hybrid quality of abstraction in terms that recall the eighth poem of 'Credences of

Summer', describing it as 'Invisible or visible or both: / A seeing and unseeing in the eye' (ibid.: 333). The weather seems to exemplify this doubleness in the concluding stanza of poem VI, as a manifestation whose terms and status shift: 'The weather and the giant of the weather' reprises unqualified conditions – that is, 'the weather' as an unadorned noun – as a distorted anthropomorphism – namely, 'the giant of the weather' (ibid.). The continuous reformulation accumulates in the following line, before the final line suggests that the weather is 'An abstraction blooded, as a man by thought' (ibid.). But there is a slippage in this metaphor, because to be blooded by thought would create the bodily world from the mind, the physical from the mental, rather than produce the mind through neurochemistry, that is, the mental from the physical. Stevens's reversed formulation shows how contingent our sense of the body is on our imagination. Transposing this relation into 'weather', the tenor of the metaphor in the preceding line, reminds us that our sense of a Nature that precedes us is retrospectively 'blooded' by our own thought of it. By failing to recognize this, we believe our experience of the weather is independent of our projections; and in so doing behave as though our emissions cannot affect it, or can at least be readily distinguished from it. Our assumptions enable contemporary climate change as their phenomenal complement.

At a point between the perceiving self and the sun in Stevens's poems sits our concept of the planet itself. The sun may be visible as a distinct body, but the earth itself is not (unless one is an astronaut). The necessity for us to imagine it, and the meanings accruing with that imagination, demonstrate our implication in its construction, much as our minds construct our experience of the weather in 'It Must Be Abstract'. Timothy Clark indicates that 'the terrestiality of one's own sensorium is implicated in the affect of the image [of the earth] in profound and inextricable ways' (2013: 16).

We can see an awareness of this problematic context in the way an abstracted planetary image is framed in Stevens's 'The Planet on the Table'. In his reading of this poem, which concludes *The Song of the Earth*, Jonathan Bate seeks to make the planet particular, transcending the mediations of the text that bring it to the table in the first place. He asks us to read the poem holding in mind 'a photograph of the earth taken from space' (2000: 282) – but then we are already at another remove from the poem itself, and the poems of

Ariel it contains. Progressively abandoning poem and picture, Bate is left with an imagined planet that he asks us to think of as 'fragile, a planet of which we are a part but which we do not possess' (ibid.). His suggestion that we 'do not possess' the planet is borne out in Stevens's allowing that Ariel's poems can only reproduce 'Some lineament or character ... Of the planet' (1997 [1954]: 450) – they are sketchy or incomplete. But in the process, they are valuable as texts because they present 'Some affluence, if only half-perceived' (ibid.).

However, Bate's perverse concentration on the titular topic of the poem (planet) rather than its context (table) ignores the immediate environment with which it furnishes us, and he cannot fully account therefore for the relationship between the texts and 'the planet of which they were part'. If we are to do justice to Bate's meditation on the poem in the light of a photograph of the earth from space, we should consider the full environmental context of obtaining such an image. Timothy Morton writes that 'We become aware of the worldness of the world only in a globalizing environment in which ... satellites hover above the ionosphere ... We are becoming aware of the world at the precise moment we are "destroying" it—or at any rate globally reshaping it' (2010a: 132). I have commented on this elsewhere, that 'Even the apparent vantage point provided by, say, a satellite is not external to the world ... but an implicit part in the creation of our understanding of it, because it is an instrument of terrestrial systems of government, science, engineering, [and] communication' (Griffiths 2012: 329). Bate's discussion of the poem is valuable in that it doesn't require the energy expenditure of an orbital shot, rather, a projection of the imagination. But his reading of the poem is entirely abstract, whereas Stevens's poem also imagines the terrestrial situation that enables our capacity for abstraction.

By figuring the poet of 'The Planet on the Table' as Ariel – perhaps named for the angel as much as for Shakespeare's sprite – Stevens transfers the lyric impulse into a fictive archetype, shading the personal into the abstract as Eliot, too, does. Where Bate's reading seeks a direct apprehension of the planet, Stevens complicates it with an intervening consciousness, meaning that direct access to the physical earth is denied. In this mediation at different scales, Stevens is able to mark again the interdependency of human and solar creativity: 'his poems, although makings of his self, / Were no less makings of the sun' (1997 [1954]: 450). If we read these lines in one way, Ariel's poems invent 'his self' and the sun – they are 'makings' or versions of both of those entities. If we read them in another, the poems are the product of – made *by* – both the (lyric) self and the enabling environment for which the sun is the dominant input of energy. Indeed, in Vendler's reading, 'our poems … are products of that solar energy that makes all things come into being. Our artificial distinctions between "nature" and "art" err: in this view, art is part of nature' (1996 [1984]: 37–8). The mutuality of these readings, and of human and solar agency, gives expression to Serpil Oppermann's more general observation that 'the natural and the cultural can no longer be thought as dichotomous categories. Rather, we need to theorize them together, and analyse their complex relationships in terms of their indivisibility and thus their mutual effect on one another' (Iovino and Oppermann 2012: 462–3). The movement of the poem from self to planet to sun illustrates the value of Stevens's poetics to understanding climate change: it situates and implicates the self in different scales simultaneously, offering a more nuanced reading than Bate's, with its focus on the referent of the planet.

Roy Sellars negotiates between the poem's possible positions when he suggests that the 'ripe shrub' that 'writhed' (Stevens 1997 [1954]: 450) 'may indicate an environmental threat or over-heated atmosphere, presaging the extinction of life on earth' (Sellars 2010: 45). Whether or not we accept this speculation will depend on the perspective from which we read the poem, and it is here that Sellars makes his argument more telling: 'Ariel as non-human, aligned with the sun (line 7), may be indifferent' to this threat, but 'From a human perspective the stakes could hardly be higher' (Sellars 2010: 45). That is to say, we have to assume Ariel's abstract position to be conscious of the planet's scale and the jeopardy it faces, but must then to return to earth, in the figure of the poet, to appreciate what the vision means for ourselves as humans. The climate-conscious critic must be aware of an implication with planetary scale while also being seated at the table.

Future imperfect: 'It Must Change'

Our point of view is altered as much by duration as by our imagined position in space, a contingency that Stevens also recognizes: his poetics demonstrates that the imposition of human order is a fiction that is subject to the passage of time. In 'It Must Change', the second part of the 'Notes Toward a Supreme Fiction', he offers one figuration of art as a statue, which 'Changed … true flesh to an inhuman bronze' (1997 [1947]: 338). The contrast between 'true flesh' and 'inhuman bronze' elaborates on the terms of 'Credences of Summer' in suggesting that a rigid art obscures the conditions of its generation by aspiring to fixity. As in 'Credences of Summer', change is again figured in the terminology of nature; in poem IV of 'It Must Change', 'Winter and spring, cold copulars, embrace / And forth the particulars of rapture come' (1997 [1947]: 339). While this may not seem to admit human input into the process of change, we can be subtly present in the way Stevens personifies winter and spring as 'copulars', identifying a natural impulse common to humans and material phenomena. That change is pleasurable too (after a fashion), given 'the particulars of rapture'. In contrast, civilization tends to resist or deny the possibility of change by envisaging a stable or at most a cyclical Nature. Botkin asserts that 'The more technologically and legally advanced a civilization, the greater the need and desire for environmental stability, for a balance of nature' (2012: 290), a resistance to change that stands in contrast to poetry's engagement with the mutable.

Nevertheless, Stevens's life offers the potential to reconcile the two tendencies, given his four decades of employment by the Hartford Accident and Indemnity Company, where as an attorney and later as vice-president he 'reviewed surety claims … making both legal and business judgments' (Grey 1991: 16). Writing in his professional capacity in 'Insurance and Social Change', Stevens suggests 'that we may well be entering an insurance era' (1997 [1937]: 793), citing 'those European countries where social pressure has been most acute and social and political change most marked [to] indicate that, as the social mass seeks to maintain itself, it relies more and more on insurance' (ibid.: 795). That is to say, as the pace of 'social and political change' accelerates, society demands exponentially greater reassurance from insurance. Stevens advises the insurance trade that 'the more they are adapted to the changing needs of changing times … the more certain they are to endure on the existing basis' (ibid.: 796).[11] Stevens's analysis shares with his poetry an insight into adaptability; Sharpe even suggests 'both poetry and insurance could be described as pragmatic responses to a world conceived ideal-istically' (2000: 147).

A more contemporary account of the insurance sector, concerning climate change rather than social change, is provided by the organization ClimateWise, which styles itself as 'the global insurance industry's leadership group to drive action on climate change risk'. The former chair John Coomber asserts that insurance CEOs 'should aim to do "something for the future" i.e. activities that are unlikely to be a money[-]earner during their tenure but are good for the long[-]term health of the firm' (Coomber 2012). These terms echo Stevens's words on business sustainability and survival, quoted above. Crucially for a textual context, Coomber remarks that 'Inherent uncertainty means that every statement made in relation to climate change risk must be caveated, but that is not an excuse for inaction' (ibid.). By the addition of more detail in the form of a caveat, the insurer's approach resembles the climate modeller's.

Rather than seek to control or account for all possible futures, however, Stevens as a poet identifies a root of imaginative understanding common to both present and the future, just as 'The Poems of Our Climate' enact both the model and the modelling process; Robert Pogue Harrison describes this root as the 'common, antecedent matrix' of both mind and nature (1999: 665). The poet's recursive syntax caveats each of his propositions, but enacts rather than exhausts the principle 'that one would want to escape, come back' (Stevens 1997 [1942]: 179). Yet Coomber's analysis, rather than investigating the principles at work in our relation with climate, prioritizes human activity, and regards 'the challenge [as] arising from the side effect of generating fossil fuel energy, the emission of greenhouse gases and their impact on the world's climate systems' (2012). If emissions are only a 'side effect', this centres our understanding on the human practice of 'generating fossil fuel energy'. Such an unreflexive, anthropocentric approach, which fails to recognize what Timothy Clark refers to as the 'epiphenomenal' quality of human experience on earth (2013: 12), is something against which Beck also cautions us when he writes that '"side effects" do not eliminate the self-endangerment to which they point, but rather intensify it' (2009: 127).

In contrast, to imagine climate change rather than define it requires us not to think of 'climate' as simply a zone in which human effects or impacts occur, because climatic phenomena also have agency and affect our culture, which is always entangled with them. Stevens envisages this interaction of human intention and phenomenal nature in horticultural terms in the fifth poem of

'It Must Change'. While fruit trees are planted, they outlive the planter and perplex his original intentions, so that where his house has stood, eventually only a 'few limes remained' (1997 [1947]: 339). Humans no longer have a place in this locale as the collapsed house represents the ruin of a controlled, orderly environment, a motif already in evidence in 'A Postcard from the Volcano'. Although there is a hint of human inscription on the land in the intentional act of planting the trees in 'It Must Change', Stevens refers to their fruit as 'garbled green', so even the 'limes' could be read as garbled 'lines'. Change is not limited to what we as humans intend to change, but is a process to which the human cultivation and direction of nature are themselves subject. Read this way, the subtitle 'It Must Change' is not an exhortation for human beings to be drivers of change, but a reminder that we need to accommodate such change in our understanding of the world.

Botkin draws a valuable distinction in this regard:

> there are kinds of changes that are natural in that they have been part of the environment for a long enough time for species to adapt to them, and many [species] require these changes. If we take actions that lead to these kinds of changes and at rates and quantities that are natural in the sense I have just described, then these are likely to be benign. If we invent some novel change that species have not had a chance to evolve and adapt to, then those are more likely to lead to undesirable results, and we should be very cautious in using them. (2012: xv)

The terms of Botkin's distinction are somewhat problematic, inasmuch as they still presume a benign or at least harmonious Nature, closer to Soper's first definition than to her second; previous extinction events were not, for instance, anthropogenic, yet were still novel enough to bring about extensive extermination across species that had no time to adapt. Nevertheless, if we are sensitive enough to limit our management of nature to the first kind of change Botkin describes, then rather than persisting in our attempts to stabilize or direct it we can more readily see where our interventions do accelerate or exacerbate such change and take it into his second category.

The few planted lime trees comprise an intervention of the first kind, because they are a change to which the landscape can readily adapt. Contemporary climate change is on the other hand 'novel' in Botkin's terms because it leads to 'undesirable results'. Moreover, its capacity to exacerbate prevailing, naturally driven change is increased because its emergence cannot be wholly traced back to the intentions, whether rational or irrational, conscious or unconscious, of human beings. At one order of magnitude, the untended lime trees represent organic growth exceeding its original function for humans, because no one remains to eat the fruit that is produced. In more general terms, this is the scenario Clark characterizes when he describes 'the immediacy of perception [as] our scalar blindness to the tree as a temporal entity, one that grows, flowers or dies etc. over a very long period of time' (2013: 11).[12] At a greater order of magnitude, climate change is not even something we can glimpse at a moment in time, as we might a tree. It persists as an intangible result of chaotic and hybridized ecological interactions between unintended human effects and environmental forces.

Climate change thus outruns the insurer's attempt to caveat it, and becomes categorically different to the 'accelerating social change' that Stevens discusses. As Nigel Clark indicates, climate change's intellectual and political difficulty entails 'not only isolating the human contribution from the "background noise" of natural climatic variability, but doing so with enough confidence to be able to apportion human forcing among geographically and historically determinate social groupings' (2010: 42). As these processes 'form a single complex global system – with its own internal dynamics and emergent properties – certain conventions of isolating specific causal agents and accounting for their contribution to overall change need to be fundamentally rethought' (ibid.: 44). By continually entangling human perception with perceived phenomena, and attesting to the agency of both, Stevens makes poetry a fuller engagement with the world than the lawyer's or insurer's attempts to discriminate and control it.

Recognizing the need for his supreme fiction to 'change' to accommodate the mutable world, Stevens situates himself in a tradition of English literary thought, in which poets have identified that their work will respond to future contexts. For instance, new works reorder our reading of the canon in Eliot's 'Tradition and the Individual Talent' (1975 [1919]: 37–44), while Percy Bysshe Shelley characterizes poets in 'A Defence of Poetry' as 'the mirrors of the gigantic shadows which futurity casts upon the present' (1977 [1821]: 508). Having examined Stevens's accommodation of prospective change in his poetics, I now consider how his conception of change is itself changed by responding to contemporary environmental understanding.

In his 'Anachronistic Reading' of Stevens's poem 'The Man on the Dump' (Stevens 1997 [1942]: 184–6), J. Hillis Miller proposes that, 'It is impossible to read [the] poem thoughtfully today without seeing how [the] dump with its single human presence anticipates our present condition' (Miller 2010: 83). While 'Stevens lived in that happy time before we became aware of climate change [and] global warming', Miller writes, as we read the poet's work now we are not so fortunate (ibid.). Miller's process of anachronistic reading 'sees a text as prefiguring a future event that comes to seem what the text predicted, foresaw, or forecast' (ibid.: 82). I would refine this by suggesting that the particular resonance of 'The Man on the Dump', its response to changing conditions, lies in the attention Stevens pays to the principles of waste disposal, the material implications of which have been exacerbated by human behaviour since the poem's composition,[13] just as 'The Poems of Our Climate' identifies a discrepancy between conception and phenomena that plays out in climate change. In an aesthetic context, Stevens's remark that 'The dump is full / Of images' (Stevens 1997 [1942]: 184) marks a despair that our ways of viewing the world are past the point of meaningful use (as language and perception are suggested to be in 'A Postcard from the Volcano'). The flora and greenery that are consigned to the dump are read by Costello as evidence 'of weariness and disgust', prompted by 'how hackneyed these images have become' (2007: 170). Yet in his choice of the dump itself as image, Stevens draws attention to something that is as intrinsic to our environment as the tired natural tropes that are among the other waste accumulating there.

In its mixture of artificial and organic imagery, 'The Man on the Dump' exhibits the reciprocity of cultural and natural agency that characterizes Stevens's poetry. At one level, the poem's dump represents a specific physical environment that Stevens knew, according to his daughter's account.[14] But the artificiality of the dump as a material construct is heightened and intensified by the process of its being rendered as metaphor: the concentration of the items abstracted on the dump signifies wider human networks in time and space: 'the wrapper' among other things hints at what it was designed to contain while a box has travelled thousands of miles 'From Esthonia' (*sic*) to end up there (Stevens 1997 [1942]: 185). In that context, Stevens's 'can of pears' shares an element of economic symbolism with Prufrock's peach or Sweeney's hothouse grapes, or indeed the carnations of 'The Poems of Our Climate'.

The accretion of these items on an *ur*-dump is symbolic of everything that humanity has to keep at bay to identify itself as humanity. Ellmann's remarks on *The Waste Land* remain apposite in this context: 'the subject defines the limits of his body through the violent expulsion of its own excess' (1987: 94). At the same time, the dump is not only an object created by human action, because it has its own agency – the trash itself represents the persistence of matter beyond its function for humanity. In attending to it, Stevens recognizes what Jane Bennett in her own encounter with litter describes as 'stuff that commanded attention in its own right, as existents in excess of their association with human meanings, habits or projects' (2010: 4).[15] 'The Man on the Dump' makes an imaginative recovery of waste that disabuses us of our sense that we inhabit a civilization with no hinterland of landfill. It serves to remind us that 'our trash is not "away" in landfills but generating lively streams of chemicals and volatile winds of methane as we speak' (Bennett 2010: vii). Stevens's explicit and fictive abstraction of the Hartford dump changes through Miller's practice of anachronistic reading to shed light on the undisclosed social myth of cleanliness and progress, imaginatively rendering its material force.

The greenhouse affect: 'It Must Give Pleasure'

Stevens recognizes that nature always exceeds our definitions of it, and through this recognition we can consider the requirement for his supreme fiction to 'Give Pleasure'. The process by which phenomena resist fixity has been figured in terms of sexual pleasure in 'It Must Change': 'Winter and spring, cold copulars, embrace / And forth the particulars of rapture come' (Stevens 1997 [1947]: 339). Rather than imposing human order on these processes, poetry, by being open to new readings in the light of literal and metaphorical changing climates, participates in the same generative forces. For instance, Stevens has identified the play of the imagination with natural renewal in 'Sea Surface Full of Clouds', because the poem's cyclical quality recapitulates the waves' own 'fresh transfigurings of freshest blue' (ibid.: 85). Stevens aspires, as a poet, towards an abstract root common to imagination and phenomenon, as Harrison attests, in which neither natural nor cultural agency has priority. This capacity for invention must then be attributed to nonhuman, unintentional forces as much as to human will. Costello defines the common capacity as 'the superfluity of human and natural creativity that stimulates change' (1998: 586). The creative impulse is superfluous to our normative sense of order, but can thus engage with transformative phenomena in a way that analytical processes cannot.

The tension between analytical and imaginative impulses can be seen in the final poem of 'It Must Give Pleasure', the third section of 'Notes Toward a Supreme Fiction'. Stevens addresses the 'Fat girl, terrestrial', the 'fluent mundo' that is the world: 'You remain the more than natural figure ... the more than rational distortion' (1997 [1947]: 351). In these two separate lines, Stevens positions the rational, which we conventionally see as the province of the human, as complementary to the natural, but only inasmuch as the 'fluent mundo' – a term that itself exceeds a more denotative description such as 'changing world' – cannot be contained by either category. Furthermore, Stevens's recursive phrasing shows that the personification of the world is a fiction, because its quality changes almost immediately from 'fat' to 'terrestrial'. The world, then, has an affect that 'cannot be imagined (even ideally) as [a] person', in Bennett's words (2010: xii). In the spirit of the alternative readings that Stevens encourages, 'the more than rational distortion' can also suggest that phenomenal nature is not identical with a 'rational distortion' that we have made of it – our imposition on the world, fitting its processes to our sense of order.

Stevens marks a distinction between our capacity for perception and the conception he wishes to rid us of in poem VII of 'It Must Give Pleasure': 'To discover an order as of / A season, ... Out of nothing to have come on major weather, // It is possible, possible, possible' (1997 [1947]: 349). The tension between perception and conception is expressed in the strain of the repeated 'possible', whose reach gestures at something unattainable, as the opening clauses of 'The Snow Man' outline conditions it is impossible to fulfil. The 'order' in 'It Must Give Pleasure' VII is derived from nature 'as of / A season'; but because it is an analogy, the season is a conception we seek in the phenomenal world – something we now see is contingent on our current interglacial episode rather than an abiding, objectively verifiable presence throughout the earth's existence. Yet Stevens as much as tells us not to (pre)conceive those seasons; we are instead 'Out of nothing to have come on major weather'. Today, the discovery of 'major weather' should be read alongside Botkin's discrimination between naturally originating and artificially exacerbated change. An increased number of instances of literal 'major weather' are seen as one result of anthropogenic climate change, arising from the very unreasoned processes that shadow our use of instrumental reason. Stevens's lines in 'It Must Give Pleasure' now, therefore, offer an expression of our failure to capture the world by 'rational distortion'.

The importance of discovery over conception highlights another aspect of the pleasurable that becomes relevant to a poetics of climate change, that is, our sensory experience of being in the world. We cannot depend on our sense experience to tell us about climate change, but if climate as an abstract can be rendered as though it has sensory presence then we will more readily register it. Costello glosses Stevens's use of the term 'pleasure' by highlighting such a sensual quality, but qualifies it with an apprehension that also seems appropriate to the context of climate change: she says to 'give pleasure' means to 'make our eyes dilate, our hair stand on end' as much as 'to satisfy a need' (2007: 179). This reminds us of our bodily contingency in the world, as for instance experienced by the old man in the tower in the third poem of 'Credences of Summer'.

In 'What to Make of a Diminished Thing', Costello reads Stevens's poem 'The Plain Sense of Things' in relation both to natural excess – her quality of 'superfluity' – and our imagined place in the world: the autumn brings us to the titular state, akin to 'com[ing] to an end of the imagination' (Stevens 1997 [1954]: 428). What remains of the imagination is its essential quality, and Stevens strives towards this because the poem observes that the 'end of the imagination' must 'Itself ... be imagined' (ibid.). To achieve 'The Plain Sense of Things' is to attain a fictive state akin to that experienced by 'The Snow Man'.

Because the imagination can take pleasure in being 'superfluous' to reasoned order, it is able to account for what exceeds human conception; the literary text leaves room for its future readings. In 'The Man on the Dump', the dump remains the object of the poet's imaginative attention even while it marks civilization's attempt to rid its consciousness of those items. In 'The Plain Sense of Things', this paradox of waste becomes embodied in the word 'waste' itself. In the line 'The great pond and its waste of lilies', 'waste' can stand for desolation, as in the title of Eliot's poem. But it can also signal surfeit or excess, as Costello argues: 'the "waste" of the lilies suggests the opposite of barrenness' (1998: 586) in 'The Plain Sense of Things'. As desolation, a 'waste of lilies' is akin to the fallen leaves in the first line of the poem; as surfeit, lilies cover the pond in anticipation of the renewal of spring from winter, as in Williams's 'Spring and All'. Unlike Williams's litter, however, Stevens's 'waste' is doubly squandering and profligate, loss and abundance; the ambivalence is inherent to his poetics. In 'The Plain Sense of Things', the imagination entertains the possibility of organic renewal despite the condition of waste, whereas in 'The Man on the Dump', the condition of waste is humanly created, representing a failure to imagine that the trash possesses the vital materialism that Jane Bennett attributes to it.[16] The lilies and the dump can be distinguished by Botkin's categories of naturally accommodated and artificially induced change.

Because it results from the accumulation of waste greenhouse gases, climate change signifies the failure of imagination that I read in 'The Man on the Dump', but on a global scale, as Miller indicates in his broad definition of the category of waste. Climate change attests that we have only imagined as far as the energy or resources we have produced and used, and relegated the emissions generated to the status of 'side effects' at most. Such waste is not licensed by Stevens's identification of a root common to imaginative and natural renewal. Rather, the poet demonstrates that we have to resort to the imagination's scope to perceive material phenomena that exceed our experience and instrumental interests.

In 'The Plain Sense of Things', the human capacity to enable waste, as compared to natural waste, is signified by the greenhouse whose 'chimney is fifty years old and slants to one side'.[17] There is no possibility for natural renewal as this greenhouse depends on human intervention to restore it; yet in terms of waste, it is still subject to the same physical principles as the lilies on the pond, the principles of Soper's second definition of nature. If we consider the greenhouse's waste in terms of desolation, its poor state of repair and decrepit chimney signal a failure to contain and harness heat: that heat dissipates into the atmosphere, and the wasteful world remains cold and icy. An alarming alternative, however, comes into consideration if the 'waste' of heat is profligate, because this entails an accumulation of greenhouse gases of a different kind in the atmosphere; the greenhouse is then a model that exceeds its parameters, entangling itself with the phenomena it was in fact designed to contain.

In '*Poésie Abrutie*' (1997 [1947]: 268), whose title goes beyond pleasure to be 'Besotted Poetry',[18] the greenhouse becomes a concentrated and intensified symbol of human interaction with the climate, in the same way the dump of 'The Man on the Dump' becomes a concentrated and intensified symbol of waste disposal: the glasshouse in the later poem 'Is brighter than the sun itself' and the 'Cinerarias' it contains 'have a speaking sheen'. Recalling that the 'sun itself' in Stevens operates as a figure for reality in its broadest terms (Vendler 1996 [1984]: 34), we can read the artificial intensification of solar energy by the greenhouse here as a figure for poetry, a concentration of natural energy by the imagination. This becomes explicit in the way that the plants are given a 'speaking sheen', a linguistic, and thus human, supplement to the reflected sun. The effect of a greenhouse, as described by Taylor, is that it 'unsettles any easy divide between the "natural" and the "unnatural," replacing it with the apparent oxymoron *artificial nature*' (2013: 8; author's italics). But as a fictive intensification of the distant solar body's energy into one multifaceted image, it also brings it within immediate sensory experience. As such, it highlights both literally and figuratively the underlying natural principles that we would otherwise ignore or relegate to our unconscious, because we more intensely witness the sun's brightness and feel its heat. We already use the phrase 'greenhouse effect' to imagine in human terms the operation of the climate, but in likening the atmosphere to an artificial structure we retrospectively impose human design on it. The greenhouse figures the fictions that we need to construct to make climate amenable to human sensation: Taylor proposes that 'the fictional "greenhouse effect" operates not in spite of the division instantiated in the glass but because of it' (ibid.: 5).

The poetics of our climate

If the irruption of major weather into our systems of thought startles us, is there anything more conventionally pleasurable that a supreme fiction of the climate can offer? However dire our entanglement in climate change is, sombre doomsterism is at best a smug and at worst an off-putting rhetorical strategy. Patrick D. Murphy speculates whether a 'pessimistic … orientation produces an emotional propaganda affect that would discourage a sense of individual agency or would cast doubt on the efficacy of enacting such agency' (2015: 37), while Erik Swyngedouw goes so far as to suggest that this approach represents a 'negative desire for an apocalypse that few really believe will realize itself' (2010: 219), with the effect of evading the political implications of climate change and pressing on with capitalism re-dressed as sustainable development.

Swyngedouw thus advocates 'the construction of great new fictions that create real possibilities for constructing different socio-environmental futures' (ibid.: 228). More imaginative and more stimulating ways of accounting for our ecological implication are therefore valuable. Stevensian recapitulations of our predicament may throw up unexpected insights, supreme fictions for our future. His 'Sea Surface Full of Clouds', rather than merely being aesthetic for its own sake, actually demonstrates the recursive imagination we should employ in considering our phenomenal environment, for instance. This quality, playful rather than programmatic, compares with the approach that Nigel Clark advocates: 'extreme conditions condemn us and other creatures to experimentation and improvisation' (2010: 49).

We cannot let playfulness become synonymous with complacency; nevertheless, our agency is equally limited if we burden ourselves with a vision of definitive climate collapse – or, conversely, sustainable utopias – as truths before they happen. We may begin to 'discover' our way through by 'coming on' it, rather than by imposing 'rational distortion' and imagining that to be a great order, because forms can emerge as much as they can be devised (as I will show in Chapter 5). The scope afforded by poetic fictions enables us to entertain and explore the imaginative consequences of our environmental interactions without committing ourselves to particular visions. Abstract, changeable images of our climatic present and future can engage us by bringing those qualities into our imaginary sense experience and remind us that our survival is 'dependent on innumerable daily acts of endurance, compassion and making-do as it is on moments of high drama or break-through' (Clark 2010: 50).

Basil Bunting and Nature's Discord

Like the two poets I have already discussed, Basil Bunting engages with the challenge of making nonhuman phenomenal agency manifest in his work. Where Eliot breaks down conventional boundaries between civilization and what enables it and Stevens seeks a creative impulse common to the imagination and the phenomenal that would make new sense of the world – as well as demonstrating the difficulty we face in projecting on to the climate an instrumental idea of order – Bunting's poetry explores that reopened relationship, embodying natural phenomena both in imagery and aurally in the materiality of his language. Harriet Tarlo observes that Bunting 'uses musical forms and terms to explore the changing, shape-shifting environment … musical elements and natural elements correspond and coexist in such a way that neither can be said clearly to be a metaphor of the other' (2000: 158). This technique enables Bunting to weave together the agencies of culture and nature in his poetry in a manner that acknowledges, as in Latour, their mutual role in creating the wold. He scales up from individual experience through the levels of region and civilization to nonhuman terrestrial and cosmic forces. The 'patrolled bounds' between our usual distinctions of scale are, in his work, zones that species can 'slither' across, such as the slowworm and rat in his long poem *Briggflatts* (Bunting 2016 [1965]: 37–61; 51; part I, lines 2–4). Equally unbounded is the selfhood that Bunting develops in his poetry. This self cannot be readily identified with the body, which is shown to be situated in and constituted by the environment throughout his work, and this transition between human and natural creations can be illuminated by theory that reads for such material agencies.

The music of Bunting's poetry attunes our ears to nonhuman phenomena but without seeking to bring us into imagined harmony with them. His poetics is informed by both a Romantic and a Modernist heritage, and he formulates an open style to express the entanglement between human cultural and wild natural agency, exemplifying Latour's quality of hybridity. This entanglement has an open-ended rather than teleological pattern in his work, and where Stevens is recursive, Bunting reaches outwards into the climate, as can be seen in his arrangement of elemental imagery such as fire and water. These patterns are complicated and disrupted by human presence, which as in Stevens' work tries to impose a direction on material phenomena that those phenomena resist, generating waste and exacerbating a tendency towards entropy. Bunting's awareness of this disruption is marked by a departure from usual narratives of selfhood, and identity is complicated by the environment of his poems. Reciprocal attempts to bring the world into harmony are thus fundamentally compromised, and we as humans must become reconciled to the process of decay.

This chapter will focus on *Briggflatts*, a poem that represents the culmination of Bunting's work and which he dubs 'An Autobiography' (ibid.: 39): it proceeds from his boyhood landscape of north-east England to London, the Mediterranean and into the mythic history of Alexander the Great, before returning to his home soil again. I will also consider Bunting's earlier work by looking critically back through *Briggflatts* to demonstrate how its vision is developed across his writing. His work allows us to envisage human beings as part of ecological co-creation; he represents our entanglement with nonhuman process without a sense that intentional direction can be achieved on either part, despite humanity's presumption of control. Such a poetics, I contend, is valuable in the articulation of human implication in contemporary climate change.

Nature in Bunting's Romantic Modernism

In light of his engagement with our experience of nature, Bunting's work has been more obviously amenable to ecocritical readings than Eliot's and, to some extent, than Stevens's. However, rather than focus on Bunting's representation of his native environment, as other critics have, I read his work here as articulating a particular relation with natural phenomena, a relation that persists into the twenty-first century even as material nature is irrevocably altered by anthropogenic climate change.

Bunting identifies and establishes this relation through a synthesis of Poundian and Wordsworthian principles. In the preface to the 1968 edition of his *Collected Poems*, he declares: 'If ever I learned the trick of it [i.e. poetry], it was mostly from poets long dead whose names are obvious' (2016 [1968]: 554). Wordsworth is the first of these, and Bunting concludes his list with his contemporaries Ezra Pound and Louis Zukofsky. The influence of both Wordsworth and Pound in *Briggflatts* is evident to Burton Hatlen, who considers the provenance of the 'concrete particular' in Bunting's imagery, for instance:

> The power of poetic language to render up such presences steadily increases as we move from Wordsworth to Pound. Too often Wordsworth ... give[s] us abstractions rather than images. But Pound's verse is full of the things of this world, perceived with astonishing precision; and *Briggflatts* is a post-Imagist poem as well as a bardic poem, each detail finely drawn. (2000: 60)[1]

In affirming this combined heritage for Bunting's work, I also contend that he disavows certain egotistical projections of Wordsworthian Romanticism as well as the self-assertions of Pound's Modernism. He enables an understanding of human–natural relations as mutually creative and influential.

For Wordsworth, the infinite reach of the imagination contrasts with the irruption of reality. When the 'soulless image' of Mont Blanc 'usurped upon a living thought / That never more could be' (1979 [1850]: 213; part VI, lines 527–8) in *The Prelude*, Wordsworth shows that his imagination outruns the world, and is brought down by the mismatch with it. But the power of the imagination persists, and Wordsworth later calls it 'That awful Power' that 'rose from the mind's abyss' (ibid.: 217; part VI, line 594), so it reascends despite the mountain's earlier 'usurpation'; this is the process of Stevens's poetics, too, but in Wordsworth's verse the imagination cannot be controlled whereas in the latter's it is the agent that repeatedly attempts to manage a world uncontrollable at the level of the quotidian, let alone the sublime.

In contrast to this explicit engagement with the imagination, Bunting's Modernism is influenced by Pound's imagistic phase, and of his contemporary William Carlos Williams's Objectivist dictum, as expressed in 'A Sort of a Song', 'Compose. (No ideas / but in things) Invent!' (2000 [1944]: vol. II,

55). Yet a truly Modernist poetics of the discrete image, the kind Pound and Williams propound, would, if pursued to its conclusion, preclude relation with the outside world. In the Objectivist arrangement, the world becomes irremediably other, brought into the text but standing for nothing other than itself. Eliot grapples with this resistance of the objective world – Prufrock's failures of self-assertion reveal the vulnerability of ego to forces it would exclude or control, as *The Waste Land* also exposes the failure of civilization's attempt to impose order on nonhuman processes. If this recognition comes only after resistance to those forces has been attempted, Bunting instead accepts and imaginatively traces our entanglement with the world, pragmatically developing a poetics from this premise. The objective world remains no less present in his work than in Eliot's, but its effect is to efface rather than fracture a sense of self.

At the opening of *Briggflatts*, the figure of a stonemason inscribes human language materially into the world and demonstrates the give and take that an acceptance of the world's objective presence entails. He exists according to natural rhythm rather than that of the clock,[2] 'tim[ing] his mallet / to a lark's twitter' and acknowledges that his mineral medium is materially responsive rather than purely passive, 'listening while the marble rests' (Bunting 2016 [1965]: 41; part I, lines 14–16). While the gravestone he makes is a signifier of human particularity, it stands in contrast to the dead man 'In the grave's slot', thus 'naming none, / a man abolished' (ibid.: lines 25, 21–2). The two processes, death and commemoration, are not opposed but inextricably entangled, because it is the dead man's decomposition that prompts the mason's composition in the stone. As in Stevens's 'The Plain Sense of Things' (or indeed Williams's 'Spring and All'), processes of organic waste are associated with renewal because 'Decay thrusts the blade' and 'wheat stands in excrement' (ibid.: 41; part I, lines 27–8). In *Briggflatts*, however, this process has none of the aesthetic vitality of Stevens's dialectic between imagination and reality, as even the birdsong heard by the mason is effortful: 'Painful lark, labouring to rise!' (ibid.: line 23). The notion of birdsong, which uses the terms of human music to describe a nonhuman source of sound, here signifies both a human acceptance of natural rhythm, and that rhythm's struggle against inertia.

Bunting's expression of human entanglement in the world, and the world's material resistance to human beings – the marble resists the mason, while the process of decay is not arrested by the erection of a gravestone – can be illuminated by Nancy Tuana's understanding of material relations as exhibiting a quality she calls viscous porosity. She writes: 'Attention to the *porosity* of interactions helps to undermine the notion that distinctions, as important as they might be in particular contexts, signify a natural or unchanging boundary, a natural kind. At the same time, "viscosity" retains an emphasis on resistance to changing form' (2008: 194; author's italics). Bunting recognizes these qualities in his admission of nonhuman material agency into the poem: the conclusion of the first section, for instance, with the lines 'Name and date / split in soft slate / a few months obliterate' (Bunting 2016 [1965]: 44; part I, lines 154–6) paradoxically 'entunes' in its repeated rhymes a process of symbolic decay, where both the solidity of the gravestone and the structure of the verse resist but are subject to elemental erosion. At the same time, the rhymes pattern an epigrammatic acceptance of what Tarlo refers to as 'the ultimate "fact" of nature' (2000: 158). As such, Bunting's poetics accepts what for Wordsworth is expressed in the troubling presence of the mountain, or indeed the quality that terrifies Eliot in the lilacs' vitality at the beginning of *The Waste Land*.

Our ease of access to troubling material phenomena in Bunting's poetry is complicated, as it is in Eliot's, when human activity seeks to contain or suppress those phenomena. A transition away from grounded understanding of the world is doubly inscribed in *Briggflatts* with the movement from the first to the second part of the poem: first, there is a change of setting from rural to urban landscape; second, and just as significant, is the change of mindset that this relocation prompts. In the first part of the poem we are exhorted to 'trace / lark, mallet, / becks, flocks / and axe knocks' (Bunting 2016 [1965]: 44; part I, lines 140–3). This list mingles processes natural and human: birdsong (or flight), carving, the run of a stream, agriculture and history. The associations 'traced' between the two spheres are reinforced by the common brevity of the Anglo-Saxon diction and the material consonance in the '-k' and '-ks' sounds. As Tarlo points out, 'the activity of nature is not so much invoked as embodied' by Bunting's sonic patterning in the poem (2000: 156).

By the second movement of *Briggflatts*, however, the poet-figure, now a young man in London, is 'a spy', and what in his boyhood was 'tracing' now serves more functional ends: he 'gauges', 'decodes', 'scans' (Bunting 2016

[1965]: 45; part II, lines 7–14). His relation with the world becomes instrumental, rather than fully sensory, although by enjambing their objects, Bunting suggests that natural phenomena still outrun human verbs of interpretation. As in the penultimate stanza of the first section, these images also entangle the natural and artificial or functional, with 'a Flemish horse / hauling beer' for instance (ibid.: lines 8–9). The weather remains present in the city, too, in the form of 'thunder', and even the human channelling of water and gas, in 'pipes clanking', registers material resistance through their sound (ibid.: lines 14–15). The pipe, a connective device, suggests a link between urban emplacement and natural resources, so rather than straightforwardly lamenting physical separation from a rural idyll, part II of *Briggflatts* is marked by its troubled reaction to nature's material excess, which cannot be entirely suppressed or instrumentalized even in the city.

This is a concern Bunting also explores in a long poem of 1935, *The Well of Lycopolis*, imagining a contemporary London that is later the subject of the second part of *Briggflatts*. On one hand, the city of this earlier poem is characterized by its absolute distinction from nature: 'The nights are not fresh / between High Holborn and the Euston Road / nor the days bright even in summer / nor the grass of the squares green' (Bunting 2016 [1935]: 22; part II, lines 33–6). On the other hand, the invocation of natural time in these lines conjures the very processes the metropolis occludes. The titular 'Well' is itself an image of the inability to mark a distinction between human signification and natural phenomena, as it represents both civilization's dependence on natural resources and water's own agency. According to legend, if a woman drank water from the Well of Lycopolis it would cause her to break her hymen, inscribing her body as though she had lost her virginity whether or not she had done so.[3]

While at first seeming cut off from nature, the realm of the human is in fact subject to nonhuman agency; Alaimo observes that 'the human body is never a rigidly enclosed, protected entity, but is vulnerable to the substances and flows of its environments' (2010: 28), a microcosm of the paranoid civilization depicted in *The Waste Land*. Bunting's line 'We have laid on Lycopolis water' (2016 [1935]: 22, part II, line 32) can therefore be read as reflecting both cultural and natural agency. It is a declaration of hospitality, when water has been provided for us, but 'laid' also has a sexual connotation in the context of the myth. As in my reading of *The Waste Land*, Bunting indicates that we cannot make an instrumental distinction between kinds of water; neither can the city distinguish itself from the natural resources it abstracts. More broadly, in a poem that, like *The Waste Land*, sardonically juxtaposes myth and modernity and which Howarth characterizes as 'satiris[ing] Bloomsbury's incestuous mixture of modernism and literary journalism' (2012: 211), the apocalyptic mode of Eliot's writing is itself subject to Bunting's pastiche.[4] The parodic tone of *The Well of Lycopolis* suggests once more that Bunting emphasizes the resistance of material phenomena to human control, whereas Eliot in *The Waste Land* sets them in tension. With the juxtaposition of civilized practice against its long-range environmental associations, Bunting's writing offers an emergent mode for dealing with concerns that human practice has aggravated in the time since he was writing.

In both *Briggflatts* and *The Well of Lycopolis*, Bunting is aware that urban environments require us to find a different language to engage with the natural, rather than simply polarizing a pure nature and a polluted culture. In that awareness, Bunting confronts a problem also faced by his Romantic predecessor. Robert Pogue Harrison writes that 'It is in the city that Wordsworth recollects the scene of nature, and it is only by recollecting his recollection that he relates to the presence of nature … The nostalgia, in turn, is what draws nature into its presence' (1993: 163). The contrast between the first and second movements of *Briggflatts*, while it is a contrast between the embedded language of youth and the self-conscious language of the metropolis, is not a shift from 'natural' language to a more alienated vocabulary. Rather, it is a movement from a language that accepts both its limitations and the counter-agency of material phenomena – as in 'The Poems of Our Climate' – to one which attempts to master their dynamic force. What at first seem straightforward and then self-important uses of language, in the first and second sections respectively, are actually both retrospective, as in Harrison's account of Wordsworth, being seen from the reflective position of *Briggflatts'* final movement. Boyhood and literary apprenticeship take place in the broader environment of that view, which is able to entertain the material value of the world in its own right, 'the loveliness of things overlooked or despised' as Bunting suggests in his *A Note on Briggflatts* (2009 [1989]: 40). These are the same things that Stevens finds on the dump, and, more extensively, that make up David Jones's anathemata, which I will explore in the subsequent chapter.

Because such material phenomena are a constant presence in *Briggflatts*, we should be wary of suggesting that those in a rural landscape are somehow more authentic than those in the city. Hatlen points out that *Briggflatts* 'is full of references to—and invokes by their Northumbrian names—the flora, the fauna, the topography, and the agricultural and domestic traditions of a specific region of the earth' (2000: 52). We see this, for instance, in the closely observed account of the mason's work, and its relation to natural time, or the discriminating ear that picks out the 'sweet tenor bull' (Bunting 2016 [1965]: 41; part I, line 1). Yet this attention to detail is not unique to the rustic situation of the first section, and persists in the city, where a gaze takes in 'lines of a Flemish horse / ... the angle, obtuse, / a slut's blouse draws on her chest' (ibid.: 45; part II, lines 8–10). We ought not, then, make the poem speak solely to regionalist agendas.[5]

Briggflatts both resists and then moves on from finding rural and metropolitan accounts of the world sufficient; and in so doing shows that Bunting is already conscious of the significance of elsewhere. He is both local and global in his outlook at the same time, hence the 'Regionalism and Internationalism' Hatlen stresses in the title of his analysis. The London of *The Well of Lycopolis* is situated in the broader mythic–historical and geographical context suggested by the title, while *Briggflatts* begins with the local, the river 'Rawthey's madrigal', and flows down to the 'strong song' of the sea by its end (Bunting 2016 [1965]: 41; part I, line 2; 61, Coda, line 1), admitting the pull of oceanic processes. Hatlen elaborates: '*Briggflatts* issues from and seeks to speak for a place that looks east to the North Sea as much as or more than it looks south to London' (2000: 54). Bunting's practice, drawn to the sea and further onwards in the manner that 'the pilot turns from the wake' (Bunting 2016 [1965]: 45, part II, line 37), thus moves in the opposite direction to Wordsworth's nostalgic impulse, which Harrison writes, 'is against the current of this river ... from the alienating openness of the sublime ... to the intimate enclosure of [its] origins' (1993: 163). Bunting allows himself to be subject to the tides, carried not home but out into the elements.

An economy of elements, the poetics of entropy

The way *Briggflatts* is drawn on into the sea is indicative of the environmental boundaries that remain open throughout Bunting's work. His poetry moves outwards from sites of human significance into elemental, material processes. *Briggflatts*, for instance, resonates with images of elemental water and fire, and intimates the consequences that instrumental uses of them have for the cycles of which they are part. Anthony Mellors can help us situate Bunting's vision in its broader literary context:

> Mythic consciousness attests to cosmic powers that allow 'man' to recover and participate in natural processes rather than symbolising the division between human significance and a chaotic universe. The shift is towards an ecological theory of artistic enactment: man is created by his environment, therefore he must learn to express himself through it, to permit himself to be expressed by it, instead of trying to beat it into shape. (2005: 23)

Mellors's crucial observation is that, once environmental factors are accepted as an influence, we cannot then look to control our nonhuman environment in a straightforward fashion, to 'beat it into shape'. Bunting expresses himself by weaving what Phillip Brown calls 'the elemental threads' of *Briggflatts* 'into a rich fabric, the complexities of the soil cycle and the water cycle inter-lacing to create an intricate pattern' (2001: 3).[6] In being open to such forces, Bunting's work is valuable to a poetics of climate change, which demands we attend to the force and agency of phenomena other than those we can inten-tionally direct.

Bunting's purpose in patterning nature is to render processes that operate at a larger scale sensible at the experiential level. This represents his engagement with a Romantic dilemma articulated by Kate Soper: 'Speech or writing mediates, either deliberately or as an effect, that which is immediate and preconceptual, and thus renders conceptual—and in the process in some sense "betrays"—that which is as it is, and is experienced as it is, only because it cannot be spoken' (2011: 21). Bunting negotiates this dilemma by working in the opposite direction, rendering process – which already has to be imagined because it takes place through time rather than instantaneously – in the movement of the verse. Through this movement, imagery such as the rocks or marble of *Briggflatts* become woven into networks of motifs rather than remaining isolated, intractable objects, as the poet mobilizes their viscous porosity. He effects an engagement with nature not solely by close scrutiny of the environment, but by giving the agency of artist and material phenomena equal status in the creation of the work of art. As a result, the processes that Bunting examines, although physical, are not reified. They are systemic rather than bounded, leading forever out of the poem to the phenomena of which they are a part. As Daniela Kato writes: 'we are left instead with a boundless unknown, a contingent flow of materials and forces – rain, earth, wind and sea – that we are nevertheless exhorted to keep on following' (2014: 166).

In ecosystemic terms, a productive analogy can be made between Bunting's vision and Beth Dempster's notion of 'boundarylessness'. She argues that ecosystems have tended to be considered bounded or 'autopoietic', that is self-generating, according to some supposedly inherent or autonomous design: her own training as a forester, for instance, 'promote[d] interpretation of forest ecosystems as autopoietic', she writes (2007: 105), objectifying them rather than allowing for their reciprocal agency. As such, forests as systems have become regarded as mechanisms amenable to human control and market management.[7] An alternative conception might be more helpful, she proposes. Contemplating the shoreline and snowline of a West Coast rainforest as an example, she argues that 'While these may be boundaries, they are not "self"-produced … the tree-line is a result of biological as well as climatic factors. If the latter are included as part of the system producing the boundary, then the boundaries must be drawn to incorporate these components, which would include a greater spatial extent, moving the boundaries further out' (Dempster 2007: 104). Any influence coming to bear on a system, such as 'climatic factors', then, must merit consideration as part of that system, even if it lies beyond geographically conceived or individually experienced boundaries. Dempster dubs such an understanding of systems as 'sympoietic' rather than autopoietic. This is akin to material ecocriticism's conception of agency as not necessarily entailing intentionality, with organisms and environments to be regarded as sites where fields of force come to bear rather than as bounded entities.

Bunting's poetics suggests that the conventionally separate zones of human and natural creativity are transgressed, in this sympoietic fashion. In a passage from part II of *Briggflatts*, the animal world is considered in artistic, organizational terms, each creature imposing a pattern on its environment. Among these, sea anemones 'design the [rock] pool / to their grouping' while specifically resisting human aesthetics, 'spit[ing] cullers of ornament' (2016 [1965]: 49; part II, lines 162–4). In doing so, they exhibit a power usually recognized in humans alone, autonomous entities extending their agency beyond their individual bodies. In sympoietic terms, the power of creation and organization is not confined to life, either – agency is ascribed to meteorological phenomena when 'Mist sets lace of frost / on rock for the tide to mangle' (ibid.: 58; part V, lines 21–2). Bunting's image demonstrates Tuana's theory of '*emergent interplay*, which precludes a sharp divide between the biological and cultural' (2008: 189; author's italics) in the production of phenomena: the 'lace' becomes a decorative overlay on the rock, but is a temporary inscription, as subject to change as the gravestone was in the first part of the poem. Hence the system is not in aesthetic harmony, but is one where the 'tide' can disorder delicate patterning of 'mist' and 'frost', all different states of water. In Dempster's terms, this is not an autopoietic system – if it were, it would exhibit 'Homeostatic balance'; instead, it is a sympoietic one, 'Balance[d] by dynamic tension' (2007: 103) between vapour, solid and liquid – mist, frost and tide.

This dynamic tension can be seen in the way the poet puts human and nonhuman agency on an equal footing in *Briggflatts*, resulting in what Kato calls 'constant traffic between the terrains of the imagination and the physical environment, the mental and the material' (2014: 166). Where Stevens searches for the creative impulse common to the natural world and the aesthetic imagination, Bunting draws imagery from the former to communicate, synaesthetically, the quality of the latter. In the fourth movement of the poem, art is again figured by the environmental as we are asked to 'consider' the music of baroque composer Domenico Scarlatti: 'stars and lakes / echo him and the copse drums out his measure' (Bunting 2016 [1965]: 56; part IV, lines 45–9). In the transition from the verb 'echo' to 'drums', the metaphor moves from a passive to an active nature, as the trumpet in Stevens's 'Credences of Summer' VIII first follows then precedes the weather it announces. Bunting's practice in *Briggflatts*, 'pragmatically admitting nature into culture's ken and vice versa' according to Sara R. Greaves, means that 'culture and nature, the visual and the auditory, tenor and vehicle, weave in and out defining each other' (2005: 69). That is to say, human presence in Bunting's poems is not the only source of creativity.

But mutual agency is not always harmonic in Bunting, as both human design and natural process can be equally destructive. In the second section of *Briggflatts*, as the scene moves to Italy, Alpine marble is shown vulnerable to both weather and industry: 'Ice and wedge / split it or well-measured cordite shots, / while paraffin pistons rap, saws rip' (Bunting 2016 [1965]: 47; part II, lines 99–101). Having established a common destructive capacity, Bunting goes on to compare the consequences of natural and industrial processes, in the image 'clouds echo marble middens' (ibid.: line 103). Water vapour that has been cast off from land and sea is made, poetically, to resemble the cast-offs of marble extraction in a parallel between a natural system and an economic one. Such an analogy is common in our understanding of the nonhuman world, with the use of labels such as "producer" and "consumer" to describe organisms' role in ecosystems being particularly indicative of an economic worldview; furthermore, regarded 'As a modernized economic system, nature now becomes a corporate state, a chain of factories, an assembly line', Donald Worster explains (1985: 313).[8]

Yet the economy is not the only possible model for natural phenomena; neither can economization account for all material processes. Clouds represent one stage of a continuous hydrological cycle, whereas the middens are dumped waste, that is, the end of an economic process of resource extraction where continuous, cyclical use is not envisaged, which complicates the notion of an 'echo' between them. In the light of economics' failure to account for such wastage, Worster considers other scientific models of the environment that recognize this kind of discrepancy: 'The ecosystem of the earth, considered from the perspective of energetics, is a way-station on a river of no return' (1985: 303) – like Bunting's poetics (and unlike Wordsworth's), it drives onward, away from the wake. Worster continues:

> Energy flows through it and disappears eventually into the vast sea of space; there is no way to get back upstream ... By collecting solar energy for their own use, plants retard this entropic process; they can pass energy on to animals in repackaged or reconcentrated form—some of it at least—and the animals in turn hold it temporarily in organized availability. (Ibid.)

If we liken our economies to ecosystemic closed loops, however, we neglect what must necessarily be wasted. As a result, waste collects as we both produce it and efface it. With this doublethink, the human economy is not an harmonious replication of the natural economy, but an exacerbation of its tendency towards entropy. Bunting's image reveals the role of human activities in aggravating an innately entropic tendency.

Given that 'paraffin pistons' – human tools that run on fossil fuels – are integral here to the extraction of marble, a comparison between economic and environmental processes can be made by considering the carbon rather than the water cycle. Vegetation extracts carbon dioxide from the atmosphere and converts it into nutrition as part of its life-cycle, as it does solar energy; in contrast, human activity simply offloads excess greenhouse gases – emissions from fossil fuel combustion, a process that derives energy from the sunlight stored by ancient photosynthesis[9] – into what effectively become atmospheric middens. Bunting also associates economic activity with its cast-offs in his poem 'They Say Etna' (2016 [1933]: 165–7), when he remarks that 'Waste accumulates at compound interest' (ibid.: 167, line 82). In the poem, waste is shown not just as the result of but as essential to the processes of capitalism, because 'Capital is everything except the desert / sea, untunnelled rock, upper air' (ibid.: lines 66–7). Capital is defined by its environmental exclusions; yet it seeks continuously to acquire them, putting the conditions for its own existence in jeopardy. Through the aspiration to acquire these 'excepted' environments, Capital then broaches the territory of excess: 'Breathed air / is Capital, though not rented: / 70 million tons of solid matter / suspended in the atmosphere' (ibid.: lines 68–71). This marks one form of accounting, where the unrented air and the airborne particles that escape human commodification are enumerated. But this enumeration, continuous with the accounting of resources used, is not able to identify the point of transgression, where there is still sufficient uncapitalized environment to accommodate the externalities of industry. By suggesting that 'Waste accumulates at compound interest', though, Bunting sardonically acknowledges the already neglected costs of human activity and their potential for exponential effects.

Thus Bunting anticipates David Wood's exegesis of externality, that 'temporal externalization—dumping waste in the river of time—makes sense under more expansive conditions. But [it] makes less and less sense as the world gets smaller' (2005: 174). In 'They Say Etna', the world is considered in terms of resources abstracted from the natural processes that accommodate them, thus generating waste. This contrast is evident in the two parodic, headline-like statements in the poem's final lines that clearly demonstrate the tension between ecosystemic and economic worldviews. The declaration **'MAN IS NOT AN END-PRODUCT, / MAGGOT ASSERTS'**, reflects on the bodily decomposition that Bunting will return to in *Briggflatts*, emphasizing human materiality as a process rather than as a 'product' of markets; by more economically oriented understanding, however, **'MAN IS AN END-PRODUCT AFFIRMS / BLASPHEMOUS BOLSHEVIK'** (Bunting 2016 [1933]: 167, lines 77–8, 83–4; author's boldface and capitals).

Our failure to contain or constrain the processes of nature characterizes other elemental imagery throughout *Briggflatts*. Water resists commodification as, reduction to, or imposition of a defined state in the poem, as it does in *The Waste Land* and *The Well of Lycopolis*. When 'fog on fells' is juxtaposed with 'spring's ending' (Bunting 2016 [1965]: 44; part I, lines 149–50), it is not just that the supposed end of a season is smudged by vaporous water, it is that the transition erases any certain seasonal boundary. The motif of unseasonal weather recurs throughout the poem, with the 'bogged orchard' and the 'damp' that 'hush[es] the hive' in the 'disappointed July' (ibid.: 49; part II, lines 172–6), or, conversely, the unexpectedly melting ice that opens the wintry fifth movement of the poem, 'Drip – icicle's gone' (ibid.: 58; part V, line 1). This has been prefigured by *The Well of Lycopolis*; passages of autumnal retrospection such as 'Scamped spring, squandered summer, / grain, husk, stem and stubble / mildewed; mawkish dough and sour bread' (Bunting 2016 [1935]: 23; part III, lines 28–30) show, as the earlier examples I have cited from the poem, the difficulty of synchronizing human time with seasonality, meaning that the agricultural efforts to put nature to use end up wasted. Meanwhile, the question 'What reply will a / June hailstorm countenance?' (ibid.: 21; part II, lines 7–8) signals both the unmanageable quality of water as unexpected weather and nature as an unanswerable agent. In these moments, seasonality is revealed to be contingent on the human imposition of order on the world, an order that natural phenomena materially resist.

Water runs through *Briggflatts* in a way that not only resists but also erases human inscription on the world. 'Rain rinses the road' in the poet-figure's native countryside, and once he is at sea, 'Fathoms dull the dale' (Bunting 2016 [1965]: 43; part I, line 92; 46; part II, line 59). The latter, oceanic distances occlude the memory of home; but there is an intimation of swamped land here as well, the 'bogged orchard' that is to come later in part II. The transience of humanity compared with water is most evident in the lines 'Who cares to remember a name cut in ice / or be remembered? / Wind writes in foam on the sea' (ibid.: 46; part II, lines 52–4).[10] Water serves a similar function in the poem's first movement when the boy's young love brings 'Rainwater from the butt ... to wash him inch by inch' (ibid.: 43; part I, lines 109–11). The cleansing process begins an act of self-erasure by the poet–narrator that continues throughout the poem, alongside the 'rinsed road' becoming 'dulled dale'. An intimate identification with the landscape is also indicated by the sly reference to the boy's testicles as 'pebbles', making stones of organs that should be seed-bearing. Whether we take the geological or procreative association from the innuendo, it situates the narrator in his spatial or historical environment, taking him out of himself into the landscape or into generational time.

There is a similar interdependence of elemental agency with the human in *Briggflatts'* motifs of fire and the poem demonstrates its dual quality of productivity and destruction when it is put to use. In the third part of the poem, for instance, we are reminded that this process can run out of control as Alexander's army makes its progress; their 'torches straggle / seeking charred hearths / to define a road' (ibid.: 51; part III, lines 5–7). The soldiers both create and follow a trail of destruction, the 'charred hearths' employing aural and imagistic concision to signify a connection between domestic warmth and unchecked combustion; Bunting's use of the elemental image flickers between its associations with utility and danger. He exploits the hybrid quality that Nigel Clark identifies in his account of fire's emergence and its adoption by humans: 'almost everywhere there is natural fire' on earth, 'there are or have been humans willing to augment the planet's own pyrophytic tendencies' – that is, its suitability for fire (2012: 269).

This augmentation takes us to the point of 'contemporary excess of anthropic combustion' that Clark implicates in climate change (ibid.: 268). Observing 'that the interplay of biological life and terrestrial fire holds the earth's atmosphere at a point which is far from equilibrium', Clark wryly concludes that now 'might not be [a] good time to risk radically supplementing the earth's combustive budget' (ibid.: 272–3). Whether we read fire as having been used for heating or slaughter in the image of 'charred hearths', we are now conscious that its waste, carbon dioxide, also accumulates at compound interest.

Bunting's form allows him to develop the significance of his elemental motifs by having them recur, as though musically, throughout *Briggflatts* to expand their resonance. In the poem's final movement, Bunting describes as 'Furthest, fairest, things, stars, free of our humbug' (2016 [1965]: 60; part V, line 86), but just as experience of the natural environment is impossible to communicate without language, these vast, stellar processes are still defined in human terms. We can thus share Brown's observation that, 'As rock and water undergo transformation, so too does fire, and the several hearths of *Briggflatts* are subsumed into the flames of the cosmos' (2001: 12). To describe a star as 'wrapt in emphatic fire roaring out to a black flue' (Bunting 2016 [1965]: 60; part V, line 88) is to invoke domestic processes of combustion to convey the stellar. Moreover, whatever the quantity of fuel remaining, the terms of comparison reveal that starfire is a finite resource. Bunting's metaphor emphasizes a sense of entropy on the cosmic scale, and as we read it today, its fossil-fuelled vehicle becomes as resonant as its tenor.

The star's light is further figured as the 'tremulous thread spun in the hurricane / spider floss on my cheek, light from the zenith' (Bunting 2016 [1965]: 60; part V, lines 93–4); as 'spider floss', the image explicitly connects across natural, personal and cosmic scales. These scales are invoked in time as well as space: 'Each spark trills on a tone beyond chronological compass, / yet in a sextant's bubble present and firm / places a surveyor's stone or steadies a tiller' (ibid.: part V, lines 89–91). Thus, although a star's existence outruns anthropocentric timescales, that is, our 'chronological compass', it helps situate human presence in the world by directing our navigation and mapping. The agency of the cosmic in the human world reminds us that we depend on processes that are orders of magnitude beyond our lives – processes over which we have no influence, and which may seem timeless but are themselves also bound by an inevitable entropy.

Bunting unbound

Bunting enables even the 'Furthest, fairest' actors to participate in human experience while they remain 'free of our humbug', demonstrating the openness of his vision to phenomenal forces. The cosmic is the sublime – yet not truly the Wordsworthian sublime, because the key expression of the Romantic poet's relation with nature involves a distinction from it. Early in *The Prelude*, Wordsworth creates a boundary between himself and his environment even at the moment he seems most responsive to it: 'For I, methought, while the sweet breath of heaven / Was blowing on my body, felt within / A correspondent breeze, that gently moved / With quickening virtue' (1979 [1850]: 31; part I, lines 33–6). Wordsworth's imagination is here enlivened in a manner akin to electromagnetic induction rather than by a literal inspiration of the breeze through nose or mouth, meaning that the meteorological and metaphorical breezes are, unexpectedly, separated at the skin. For Patricia Waugh, the passage marks a recognition that the form of the aesthetic is found in the world: 'In Wordsworth's writing, ... we can detect that form situated in nature, a blessing in the gentle breeze which actually blows upon us to meet a corresponding breeze within' (1992: 22), with the wind inspiring the shape of its imaginative reception. However, this correspondence between the breezes is not a transaction across boundaries as such: the human *in* the environment is still separate *from* it. In Bunting's work, a dispersal of selfhood and what it means to be human addresses our implication in the environment in a way Wordsworth's internalized process do not; as Mellors points out, a Late Modernist poet instead 'learn[s] to express himself through it, to permit himself to be expressed by it' (2005: 23).

Bunting modifies the Romantic discovery of the self in nature to create a sense of permeable, contingent identity, conditioned by and conditioning its environment. The elemental patterning of water in *Briggflatts* resists reduction to a single significance and erases human traces, though its permeation of selfhood is a motif Bunting has developed from his earliest work onwards. The third ode of Bunting's *First Book of Odes* for instance opens with a first-person expression of passionate intensity, 'I am agog for foam', but quickly loses this 'I' in a profusion of plural pronouns 'our loneliness ... our envy', 'Its indifference / haunts us' and so on (2016 [1926]: 79, lines 1, 11–14). Individuality here is expanded into a shared identity, and any character this has is contingent on the elements. Bunting opens 'the possibility of relinquishing boundaries' (Dempster 2007: 97), both between individual humans, and between culture and nature. Compare this with the opening of *The Prelude*: Wordsworth's 'blessing in this gentle breeze' (1979 [1850]: 29; part I, line 1) is instrumental in the creation of the self with the sympathetic 'corresponding' mild creative breeze it engenders. In Bunting, while the environment conditions human experience, that experience is not individuated. Neither is nature benign in the ode; it is indifferent, not even comprehensible.

The simultaneous influence and incomprehensibility of oceanic force is taken a stage further in Ode 17. In this poem, the incoming sea is resistant not only to explanation, but also defies attempts to order it, physically and linguistically. The poem opens: 'Now that sea's over that island ... I resent drowned blackthorn hedge, choked ditch, / gates breaking from rusty hinges, / the submerged copse, / *Trespassers will be prosecuted*' (Bunting 2016 [1930]: 93, lines 1–8; author's italics). This expresses a failure of imposed boundaries, the 'hedge, ... ditch, / gates', to contain the weather. Although the narrator claims to 'resent' these drowned boundaries, the tone is more resigned than bitter, with the first line implying that the rising sea has been anticipated before 'Now'. By quoting the sign '*Trespassers will be prosecuted*' in this context, the limits of cultural order are revealed, as though the sea should be subject to legal admonition. Even to identify and name a place as 'that island' is to intimate that it has always been an island and will remain so, although as Stevens points out in 'Variations on a Summer Day' these linguistic distinctions exist 'by grace alone' (Stevens 1997 [1942]: 215).

Whereas Stevens acknowledges the imaginative interdependency of poem and sun with the punning lines 'his poems, although makings of his self, / Were no less makings of the sun' (1997 [1954]: 450), Bunting signals the same contingency of text on environment by the physical failure of the sign to prevent flooding, using image rather than wordplay. The two poets present alternative answers to Timothy Morton's question about how we constitute an environment: 'At what point do we stop, if at all, drawing the line between *environment* and *non-environment*: The atmosphere? Earth's gravitational field? Earth's magnetic field, without which everything would be scorched by solar winds? The sun, without which we wouldn't be alive at all? The Galaxy?'

(2010a: 10; author's italics). Stevens reads our environment as being at least as far as the sun, while Bunting's vision is cosmic, as we have seen, setting no limit. By suggesting that it is human-imposed limits that are imaginary or inherently vulnerable to transgression, he demonstrates a boundaryless thinking akin to that described by Dempster.

Ode 17 develops through the simile 'a film of light in the water crumpled and spread / like a luminous frock on a woman walking' to the narrator's regretful 'Very likely I shall never meet her again' (Bunting 2016 [1930]: 93, lines 15–16, 30), suggesting that 'nature' in the poem may in fact simply symbolize the female. Hatlen proposes that in *Briggflatts* likewise, the 'variable' of the natural world is 'usually an extension of the domestic/erotic world of the Woman' (2000: 54).[11] In *The Song of the Earth,* on the other hand, Bate reads a reverse analogy in the first part of that poem: 'the girl he [the poet–boy] lies with is, like Wordsworth's Lucy, an embodiment of the land' (2000: 234). Greaves also draws the female/nature parallel in her analysis of *Briggflatts,* but her ecofeminist reading highlights the irreducibility of either to the other. She also points up the agency of both as opposed to the agency of (a) man: 'the most obvious Romantic Others, woman and nature, are not merely passive receivers and enhancers of the active masculine sublime, but active agents themselves' (2005: 69).

Reading Ode 17 in light of these observations, it is clear that here, also, the female and the natural are distinct, because the former fails in her attempt to control and refine the latter, to cultivate 'her garden' that she 'had prepared ... for preservation' (Bunting 2016 [1930]: 93, lines 17, 22). Her effort, conducted 'not vindictively' (ibid.: line 23), suggests that she is willing to accept natural agency rather than assert human design on it, and subsequent lines support such a reading: 'Nobody said: She is organising / these knicknacks her dislike collects / into a pattern nature will adopt and perpetuate' (ibid.: lines 25–7). This disavows human attempts to impose a 'pattern' on the world, accepting natural agency rather than attempting in vain to prevent its 'trespass', as in the first stanza.

Bunting's lines leave open the possibility, though, that we might create 'a pattern nature will adopt and perpetuate' if we do not have regard to our own boundaries. Indeed, in the form of excessive greenhouse gas emissions, we exacerbate an entropic tendency in the earth's systems, setting a pattern nature will 'perpetuate' by inducing the positive feedback cycles that accelerate warming. This interplay, in which the environment in its totality creates the human, and the human in turn shapes its terrestrial environment, means it is possible to read the self of Bunting's work as a sympoietic system,[12] which in Dempster's words is 'maintained by dynamic interdependencies' with 'neither temporal nor spatial boundaries' (2007: 94). In Bunting's poems, selfhood and civilization are contingent on natural particularities and process, and as such there can be no easy, lyric, presupposition of the self.

Where Wordsworth finds form in the world that is met by responsive forms within, that too means that the self changes and develops rather than emerges autopoietically. Isobel Armstrong writes: 'Experience is sequential and our collective analysis of it is sequential too, and the process of externalising and repossessing experience is not merely the analysis of a prior process ready to create further experience but the *subject* of a further one. Through the temporal process analysis is returned to the self as *experience*' (1982: 38; author's italics). The self is cumulative through time, and through language, in a work such as *The Prelude*. Wordsworth is not seeking to describe a self that exists discretely or completely at any stage, but developing that self through the process of poetry. In Bunting, identity is still more contingent; where Wordsworth was 'Fostered alike by beauty and by fear: / ... In that beloved Vale' (1979 [1850]: 45; part I, lines 302–4), there is no such balance in Bunting, because the environment disrupts as much as shapes the sense of selfhood.

Brian Conniff's analysis of *Briggflatts* makes such an understanding central to Bunting's poetics, especially given that the poet himself styled the work as '*An autobiography,* but not a record of fact' (Bunting 2016: 538; author's italics). Conniff characterizes the poem as anti-lyric, because, conversely, a 'timeless lyric paradise has an ultimate and coherent vision, but it has no convincing physical vision. It has idealized love, but it has no actual care' (1988: 196). By rejecting such an uninterrogated projection on to the world, Bunting, in *Briggflatts*, conveys an understanding valuable to contemporary environmentalism, namely that we can have no worthwhile ethic of care for our global environment if we prefer a mythic, idyllic Nature.

The poem's anti-lyrical tendency also enables us to see more easily its resistance to a consistent narrating first-person singular. Waugh comments that '"I" is a logical fiction necessitated by grammar and official biography

... closing down ethical possibilities of being by fixing the self in social convention and oppressive tradition' (2001: 25); by contrast, Conniff explains that in *Briggflatts*, 'One individualized voice or another is always speaking, but not one of them seems to get very close to a "record of fact" about any author, real or imagined' (1988: 182). In the opening part of the poem, the figure of the poet is a boy or a young man in the third person, objectified only inasmuch as a retrospective view is able to 'externalize and repossess' its earlier self, to paraphrase Isobel Armstrong (1982: 38). The identity of this figure is further disrupted as we enter the poem's second movement where, Conniff contends, the 'seasonal narrative structure no longer holds together, at least in the expected way, because the young poet-to-be of section one has disappeared in the shadows; he has turned into a dispersion of figures' (1988: 172). Among the roles Conniff enumerates are the 'Poet appointed', 'spy', 'pilot' and so on (Bunting 2016 [1965]: 45; part II, lines 1, 7, 37), and these come about because the poet 'finds that each culture ... necessitates a new identity, a new disguise' (Conniff 1988: 172). Kato summarizes: 'the manifold vicissitudes of the poet–hero constitute neither the centre of the poem nor a mere backdrop against which it unfolds; together with the stories that compose *Briggflatts*, they inscribe and weave themselves into its texture' (2014: 160).

In the third, central section of *Briggflatts*, the poet's own multiple, biographical figures are replaced by the voices of a soldier in Alexander the Great's army[13] and the slow worm that was introduced in the first movement of the poem. Together with the figure of the conqueror himself, the section presents three aspects of selfhood, each entailing a different relation with the environment – nostalgia, conquest and dwelling. The mobilized soldiers lament that they 'desired Macedonia, / the rocky meadows, horses, barley pancakes, / incest and familiar games' (Bunting 2016 [1965]: 52; part III, lines 72–4). Even in their longing, however, their home is not idealized, either as physical landscape or as society. In contrast, Alexander seeks to master the way ahead. His men 'deemed the peak unscaleable; but he / reached to a crack in the rock / with some scorn, resolute' to make his ascent (ibid.: lines 76–8). This marks a direct bodily engagement with the mountain that perplexed Wordsworth's imagination, despite its material resistance to Alexander's climb, 'file sharp, skinning his fingers' (ibid.: line 80). Overreaching himself, the figure of the conqueror is subsequently cast down to lie 'on glistening moss by a spring' where he hears the other voice of the poem's third part, the slow worm (ibid.: 53; part III, line 97).

Neither aspirational like the conqueror nor backward-looking like his army, the slow worm is content in its ecological niche: 'Ripe wheat is my lodging … I prosper / lying low, little concerned' (ibid.: lines 106–13). Nevertheless, like the anemones of part II, it still has an impact on its environment extending beyond its own body, as the wheat's 'swaying / copies [its] gait' (ibid.: 109–10). Mediating between the soldiers' nostalgia and Alexander's ambition, the slow worm acknowledges that presence in the world entails a relation with it and an effect that cannot be confined to the immediate environment. As the section ends, being and world are brought into literal harmony 'where every bough repeated the slow worm's song' (ibid.: 54; part III, line 130), but the movement of the poem means this is a momentary respite rather than a permanent state of stability.

The poet's own non-lyrical identity through his multiple roles, broadened in part III by the military and reptilian voices, is part of a wider pattern where human presence is fragmented through the poem. Rather than trying to assert human individuality 'into a pattern nature will adopt and perpetuate' (Bunting 2016 [1930]: 93, line 27), the unbounded selfhood of *Briggflatts* is registered in the world in other ways. Greaves places an emphasis on reading 'A Poetics of Dwelling' in the poem: 'the self is refracted through a range of personae, human and animal, as if to deny humankind its supremacy. The landscape is fused with parts of the biological body, eroticised by the dispersal in the text of sexual metaphors such as the pebbles and the slowworm, infusing it with desire' (2005: 69). The poem is not then autobiography but a natural history, the story of the environment's unshaping of a self.

One of the most striking instances of the erotic relation that Greaves identifies in *Briggflatts* is in the sexual union of Pasiphae and the 'god-bull' at the close of the second movement: 'nor did flesh flinch / distended by the brute / nor loaded spirit sink / till it had gloried in unlike creation' (Bunting 2016 [1965]: 50; part II, lines 199–203). In his own exegesis of these lines, Bunting writes: 'Those [people] fail who try to force their destiny … but those who are resolute to submit, like my version of Pasiphae, may bring something new to birth, be it only a monster' (2009 [1989]: 40). He contrasts here the failure of assertive personalities, among which we could number Alexander, with a necessary endurance of forces outside our control. The 'monster' brought to birth by Pasiphae's submission is the Minotaur, the result of interaction between the conventionally segregated spheres of the human and the natural; its image, though never explicit in the poem, is the problematic shadow of both the 'tenor bull' of the opening line and the conqueror's assertive masculinity. As an 'unlike creation', it corresponds to the category of neither man nor beast.

In *We Have Never Been Modern*, Latour suggests it is precisely the distinction between categories of human and natural that enables such hybrids, however. He argues that we as self-styled moderns 'innovate on a large scale in the production of hybrids' and that this 'is possible only because [we] steadfastly hold to the absolute dichotomy between the order of Nature and that of Society' (Latour 1993: 40). The concept of hybridity is not unique to modernity, as the Minotaur myth shows; what is uniquely modern, however, is the separation of categories that creates them in such excessive quantities. Latour goes on to ask 'where are we to classify the ozone hole story, or global warming or deforestation? Where are we to put these hybrids? Are they human? Human because they are our work. Are they natural? Natural because they are not our doing' (ibid.: 50). Bunting indicates how monstrous these hybrids are, even when we accept their presence in the world; Latour suggests further that the agency of nature remains and is exacerbated when we do not recognize or accept it. Hence, global warming is 'not our doing' in an intentional sense, even while it is the product of our deliberate practices or 'work'.

Latour describes these entangled agencies as a network, and 'the idea of the network is the Ariadne's thread of these interwoven stories', although 'the delicate networks traced by [her] little hand remain more invisible than spiderwebs' (ibid.: 3, 4). While the Daedalean labyrinth is, like the Minotaur, not explicitly referred to in *Briggflatts*, there are a number of allusions to similar structures in the poem, which mingle animal and human, natural and aesthetic agency. For instance, there is the 'rat … daring / to thread lithe and alert Schoenberg's maze' (Bunting 2016 [1965]: 49; part II, lines 181–3, 4), and the 'Tortoise deep in dust or / muzzled bear capering / [that] punctuate a text whose initial, / [is] lost in Lindisfarne plaited lines' (ibid.: 47–8; part II, lines 115–18). These are woven among numerous references to natural networks throughout the poem, such as the 'lace of frost' (ibid.: 58; part V, line 21), and the 'shadows [that] themselves are a web' in the sentence preceding Pasiphae's ravishment (ibid.: 49; part II, line 188). Kato observes that 'the poem is pervaded by all sorts of line imagery – threads, webs, tangles, paths, mazes, animals that make threads … configuring shapes of permanent movement and metamorphosis' (2014: 161). This entanglement of aesthetic and phenomenal networks into the labyrinthine structure of the poem gives expression to the complexity that anthropogenic environmental change demands we recognize. In theoretical terms, it can be seen in Alaimo's plea for a 'trans-corporeal ethics' which 'calls us to somehow find ways of navigating through the simultaneously material, economic, and cultural systems that are so harmful to the living world and yet so difficult to contest or transform' (2010: 18). As in Eliot, Bunting's use of myth presents a way of understanding phenomena that exist outside or across too-readily demarcated categories.

Pasiphae's acquiescence to the mythical bull belies benign visions of a restorative or nurturing Nature; while this denies her the active agency that Greaves claims is evident in some of the female figures in Bunting's work (2005: 69), the poet nevertheless avoids a simple identification of femininity with nature, as he also does in Ode 17. Again, his work contrasts with Lucy-like, conventional 'constructions of nature as female (as mother/virgin)' that are critiqued by ecofeminist scholars such as Gretchen Legler (1997: 228). Legler argues that it is unchallenged constructions such as these which 'are essential to the maintenance of this harmful environmental ethic and … hierarchical ways of thinking' (ibid.).[14] Bunting instead subverts hierarchy by revealing the hybrids it creates. The bull as brute nature imposes itself on Pasiphae, who, in the terms of Bunting's commentary, may serve as a representative not specifically of the female but of the human. Compare the opening of the third movement of *The Well of Lycopolis*, in which a self-consciously mythologized nature ravishes 'Infamous poetry, abject love': 'Aeolus' hand under her frock / this morning. This afternoon / Ocean licking her privities. / Every thrust of the autumn sun / cuckolding / in the green grin of late-flowering trees' (Bunting 2016 [1935]: 22; part III, lines 1–7). While both scenarios subjugate the female to the natural, the subjugating agent is not the human male, though it displays masculine attributes. Indeed, *The Well of Lycopolis*'s narrator regards the elements as having taken the woman from him, complaining 'I shall never have anything to myself' (ibid.: line 8). The implication is that we must endure and adapt to the exigencies of the callous nonhuman world, its 'green grin' as disturbing as Eliot's lilacs, rather than attempt to master these. While this valorizes a masculine kind of endurance, it does not enable mastery of the other, whether that 'other' is nature or women. Bunting's poem reminds us that we live close to such alien, uncontrollable entities. In this respect, the poet's practice can be read in terms of Timothy Morton's 'ecological thought', which situates us in a 'mesh' that 'permits no distance' and 'means confronting the fact that all beings are related to each other negatively and differentially' (2010a: 40, 39).

Mapping the order

Bunting takes on Wordsworth's legacy by finding a mesh-like weave in the world, while recognizing that human selfhood is entangled in this pattern rather than containing it. The danger of imagining that we can comprehend natural processes because we recognize this pattern is that it leads to the belief that we can master them – the same presumption of instrumentalization that perplexes climate modelling, according to Bronislaw Szerszynski (2010: 19). But for every intentional intervention asserting human order on the environment, there is an unintended effect, corresponding with the unconscious mind. Theodore Roszak suggests how both superego and id end up becoming manifest in the world: 'Precisely because we have acquired the power to work our will upon the environment, the planet has become like that blank psychiatric screen on which the neurotic unconscious projects its fantasies' (1995: 5).[15] As such, the relegation of waste to our environmental unconscious consequently leads to its accumulation and the formation of hybrid phenomena, exemplified by anthropogenic climate change.

Briggflatts, too, makes evident that the wilful imposition of order actually generates uncontainable disorder, as a result of the distinction between intentionality and unconscious. Conniff claims that 'The poet turned to the natural, objective world for a sense of order; but everything in this world ... is relentlessly active, as though it were all moving in defiance of the aesthetic desire to arrange it all, the would-be poet's will to mastery' (1988: 171). Hence the poet's eventual realization that 'he does not have to pretend that his own life ever appeared to him as a coherent story ... his world always controlled him more than he ever controlled it' (ibid.: 183). Such an unravelling of self-imposed order in the poem, which is styled as an autobiography, can be traced throughout Bunting's writing.

We have already seen how the narrator's mood rises and falls with the movement of the tide in Bunting's third ode; although Bunting's biographer Richard Burton maintains that 'the poet [is] at one with nature' (2013: 132) at the ode's beginning, there is more to the piece than the Romantic harmony this suggests. The narrator announces 'I am agog for foam' as the tide is 'Tumultuous come / with teeming sweetness to the bitter shore', and as the poem draws to its close 'we again subside / into our catalepsy' (Bunting 2016 [1926]: 79, lines 1–2, 26–7). This is not just a simple synchronization of human with tidal vicissitudes as Burton suggests, however, but an exchange across them: the sea's 'indifference' becomes 'mad waves' that 'spring ... / towards us in the angriness of love / ... tossing as they come / repeated invitations ... of unexplained desire' (ibid.: lines 18–22). The world beyond the human has an intrinsic appeal and 'The dear companionship of its elect / deepens our envy' (ibid.: lines 12–13). Yet it remains beyond us, because the sea's desire is 'unexplained', and the sky's 'endless utterance of a single blue' remains 'unphrased' (ibid.: lines 22, 5–6). That is to say, the world's affect cannot make sense in human terms even while it appeals to human senses. The effort to express it is not an imaginative struggle, as it is for Wordsworth, so much as an erasure of human identity. This ode anticipates *Briggflatts*'s 'unscarred ocean', which is not inscribed by human beings but only by another natural force: 'Wind writes in foam' (Bunting 2016 [1965]: 46; part II, lines 42, 54).

Bunting suggests, then, that language can only pattern the human experience of material phenomena rather than order those phenomena themselves. To assess what makes his poetics distinctive from Stevens's, which makes a similar claim, we can compare the ode with the more intellectual engagement of the American poet's 'The Idea of Order at Key West' (1997 [1934]: 105–6).[16] Both poems convey a proximity to and a haunted relationship with the sea, although Bunting's is more erotically charged than Stevens's. His is immediately felt, 'agog' from the beginning, whereas Stevens's poem paradoxically makes the sea meaningful by repeatedly attending to 'The meaningless plungings of water and the wind' (Stevens 1997 [1934]: 105). This insistence creates an impression of something that nevertheless *needs* to be understood, that demands our attention but is inexpressible. By the end of the poem, Stevens observes a 'rage to order words of the sea', even though those words amount to 'ghostlier demarcations, keener sounds' (ibid.: 106). In this respect, the poem acknowledges what goes beyond civilizing experience, just as *The Waste Land* contends with the intransigent nonhuman forces operating against cultural order. Knickerbocker writes of Stevens's poem that 'the sea looms through the form of the poem (what the poem *does*) even though the words seem to alienate the sea in what they *say*, and from an environmentalist standpoint this formal subversion is positive' (2012: 27; author's italics). Compare with Stevens's 'ghostlier demarcations' the sky's 'utterance of a single blue / unphrased' in Bunting (2016 [1926]: 79, lines 5–6) – something expressive but not explicitly expressed, evading understanding on human terms.

Bunting and Stevens both make the world's expression of itself syntactically conditional, with Bunting's '*If* the bright sky bore / with endless utterance of a single blue / unphrased' (ibid.: lines 4–6; my italics) comparable with, for instance, Stevens's '*If* it was only the outer voice of sky' (Stevens 1997 [1934]: 105; my italics). This conditionality acknowledges the necessary fictionality of our engagement with the world. Where human understanding is contingent, though, natural expression is in contrast 'endless' in Bunting and part of a subsequent 'summer without end' in Stevens. These are not expressions of timeless, lyrical Nature, however; rather, they represent the processes of Soper's second definition of the term, processes that are 'indifferent to our choices, will persist in the midst of environmental destruction, and will outlast the death of all planetary life' (1995: 159–60). This 'indifference' is even made explicit in Bunting.

The contrast between the two poems again comes through in their respective attributions of agency:

the sea	It was her voice that made
trembling with alteration must perfect	The sky acutest at its vanishing.
our loneliness by its hostility	
(Bunting 2016 [1926]: 79, lines 9–11)	(Stevens 1997 [1934]: 106).

In Bunting, the sea is invoked to 'perfect' one of 'our' qualities – that is, we are subject to oceanic processes beyond our control;[17] in Stevens, on the other hand, it is the singer, another self, who perfects and defines ('made ... acutest') the sky above the sea. Thus in Bunting, nature controls the human self, and in Stevens, the self orders – or attempts to order – nature, albeit aesthetically rather than instrumentally. Yet the very self-consciousness of that process in Stevens's poem – and also of the way, later, the lights in the harbour 'Mastered the night and portioned out the sea' (Stevens 1997 [1934]: 106) – shows that these absolute boundaries do not inhere in the atmosphere or hydrosphere, but are impositions from human perception, hence the wistful tone of the 'ghostlier demarcations, keener sounds' that are beyond us in his poem's final line. Bunting and Stevens recognize in their poems that the world reacts to us in ways that cannot be satisfactorily accounted for in human terms: human order deteriorates in Bunting's work and has to be recapitulated in Stevens's. Our failure as a civilization to accept this is one of the conditions that has helped engender anthropogenic climate change.

Ode 3's impassioned submission to the sea may lack the intellectual nuance of Stevens's poem. However, in 'Chomei at Toyama' (2016 [1932]: 63–74), Bunting finds a stronger voice in which to articulate human failure to order the world according to the imagination. As a free adaptation of a medieval Japanese work, the poem also takes up the challenge of a text's changing position over time, as I suggested was the case with Pound and Eliot in Chapter 2. 'Chomei at Toyama' is written in the voice of twelfth- to thirteenth-century Japanese writer Kamo no Chōmei, whom Bunting notes 'belonged to the minor nobility of Japan and held various offices in the civil service ... He retired from public life to a kind of mixture of hermitage and country cottage at Toyama on Mount Hino and there, when he was getting old, he wrote the Ho-Jo-Ki in prose, of which my poem is in the main a condensation' (ibid.: 539). In the poem, Bunting's Chōmei epigrammatically advises 'To appreciate present conditions / collate them with those of antiquity' (ibid.: 67, lines 81–2), and the lines enact this principle as an adapted restatement of it, by representing it some 700 years later. Pound has done the same in the *Cantos* by deriving texts from Homeric Greece, Confucian China, medieval Italy and the nascent United States, collating these with events of the early to mid-twentieth century implicitly in some places and more explicitly in others. In Bunting's case, the restatement of an historic principle suggests that historical change or progress is limited, because Chōmei's sentiment still pertains in the era of

Modernism. Two forms of time, cyclical and linear, are thus set in opposition, as they are in *The Waste Land*. In a letter to *Poetry* magazine's associate editor Morton Dauwen Zabel negotiating the poem's publication, Bunting commented on the pattern he sought in composing it:

> the balance of the calamities and consolations pivoted on the little central satire, the transmogrifications of the house throughout, the earth, air, fire and water, pieces, first physical then spiritual, make up an elaborate design which I've tried not to underline so that it might be felt rather than pedantically counted up. Also the old boy's superficial religion breaking down at the end needs what goes before to give it relief, and what goes before needs the breakdown to anchor it to its proper place. (1933: 1)

Later, in *Briggflatts*, a similar process occurs as the self emerges from nature, as an object of nostalgia, to end in cosmic dispersal.

Bunting's 'Chomei at Toyama' privileges the elemental, and natural disasters are for instance prominent among the events recalled by the poem's narrator: the first is a fire that destroyed 'In a night, / palace, ministries, university, parliament', then a cyclone of three years later after which 'Not a house stood' (2016 [1932]: 65–6, lines 23–4, 39). These disasters lay waste to the institutions of civilization, and as such Chōmei describes them as 'Massacre without cause' (ibid.: line 51). The phrase highlights the difficulty we have in comprehending natural agency, because Chōmei has to liken it to a human atrocity, a massacre, although a cyclone lacks the intention that such slaughter would require. It registers the agency of natural forces, but still has to conceive of them in human terms.

To understand contemporary climate change, we, too, can 'collate' it with the conditions of the past. If humans already recognized their vulnerability to natural disaster, both in Chōmei's time and Bunting's interwar writing, how much more dangerous can such material phenomena become when we presume ourselves superior to them and unconsciously exacerbate their agency? Tuana reflects on this question in considering how to talk about a contemporary meteorological catastrophe, 2005's Hurricane Katrina. She asks:

> Does it make sense to say that the warmer [sea surface] water or Katrina's power were socially produced, rendering Katrina a non-natural phenomenon? No, but *the problem is with the question*. We cannot sift through and separate what is "natural" from what is "human-induced," and the problem here is not

simply epistemic. There is scientific consensus that carbon dioxide and other greenhouse gases are raising the temperature of the Earth's atmosphere. These "natural phenomena" are the result of human activities such as fossil fuel combustion and deforestation. But these activities themselves are fueled by social beliefs and structures. (2008: 193; my italics)

Tuana continues, 'This does not mean that we cannot attempt to determine the extent to which human factors increased the intensity of a hurricane or some other weather-related phenomena' (ibid.). The weather event exposes the traces of agency. Bunting's Chōmei interrogates the significance of the cyclone with his own question: 'Portent?' (2016 [1932]: 66, line 52), and if we were to ask this same question of Katrina's significance it could be answered affirmatively, albeit tentatively. We make Katrina a site for debate about climate change – a 'portent' of it – because it is spectacularly visible in precisely the way climate is not, exposing the entanglement of agency that Tuana discusses. The weather event is productive in this way because 'You can't visualize the climate', as Morton reminds us (2010a: 28).

Within the movement of Bunting's poem, Chōmei's question specifically asks whether these natural disasters portend the 'thunderbolted change of capital, / fixed here, Kyoto, for ages' (2016 [1932]: 66, lines 53–4). As we saw in Ode 3, Bunting can ascribe human qualities of 'utterance' to natural process, but such transference is reversed here so that 'thunderbolted', which at first seems to affirm that weather necessitated Kyoto's relocation, is in fact a metaphor for the speed with which the move was carried out: human process is described in natural terms. Even though 'Nothing compelled the change nor was it an easy matter' (ibid.: line 55), the position of 'thunderbolted' so soon after fire and cyclone complicates the semantic fields of human and natural agency. Chōmei's cyclone then resembles Katrina inasmuch as it serves to reveal human entanglement with meteorological processes, though the poem is concerned with a society's vulnerability to the effects of weather rather than its complicity in their causes. Where Katrina complicates matters is that modernity's too-rigid distinction between human and natural realms is obscured by the hybrid material phenomena that result from this categorization. Tuana observes that, 'Agency … emerges out of such interactions; it is not antecedent to them. Our epistemic practices must thus be attuned to this manifold agency and emergent interplay' (2008: 196).

To find a commonality between a medieval Japanese cyclone and Hurricane Katrina is to attest to the persistent materiality of our relationship in and with the environment. The relation remains the same across the centuries despite the notion of history as progressive. Bunting was alert to the fallacy of this notion, and indeed wrote to Zabel's boss at *Poetry*, Harriet Monroe: 'The curiously detailed resemblances between mediaeval Kyoto and modern New York are not my invention, and I didn't feel called on to disguise them' (1932: 1). The poem expresses its acceptance of perpetual environmental contingency both in Chōmei's tone of resignation – 'Men are fools to invest in real estate' (Bunting 2016 [1932]: 66, line 34) – and the continued description of human catastrophe in meteorological terms – 'a thunder of houses falling' (ibid.: 67, line 104). Human and natural become corresponding threats when Chōmei writes that his new home 'stood on the flood plain. And that quarter / is also flooded with gangsters' (ibid.: 69, lines 151–2).

Nature at the end

These entangled agencies reach a culmination in *Briggflatts*, as my analysis throughout this chapter has indicated. Like Wordsworth and Chōmei, the poem looks for an apparent order in nonhuman nature through which to express a human understanding. Bunting elaborated on the pattern of *Briggflatts* in an interview with Peter Quartermain and Warren Tallman, illustrating it with the diagram seen in Figure 2.

The poet explained that, having sketched a five-part pattern for composition, the next thing was 'to look at it and … say obviously what any poet thinking of shape would say … : Spring, Summer, Autumn and Winter' (Bunting 1978: 15). Quite why it is 'obvious' that five movements would represent four seasons is unclear. What is most telling, however, is that it is the peak-scaling arrogance of a human conqueror – Alexander's assertiveness – that disrupts the seasonal order, by rising to bisect it in the central, third part of the poem. Even while seeking confirmation in one natural process – that is, the movement of the seasons – the poem is repatterned by another, the formation of a mountain.

Alexander's effort in ascending the 'unscaleable' peak brings him before the angel 'Israfel, / trumpet in hand, intent on the east' waiting for 'the signal

Figure 2 Basil Bunting's sketch of *Briggflatts'* structure (Bunting 1978: 15).

[to] come / to summon man to his clay' (Bunting 2016 [1965]: 52–3, part III, lines 76, 90–1, 94–5). Teetering on the brink of catastrophe, the conqueror tumbles to earth, waking 'on glistening moss by a spring' where he encounters the slowworm. In this creature's words, he is reminded of his place in the world, as it tells him: 'I prosper / lying low, little concerned' (ibid.: 53, lines 112–13). Alexander thus fails in his ambition, but the poem still has scope to use this as a demonstration of humility – an apocalypse in the sense of revelation rather than cataclysm.

A comparable mountain misdirection occurs for Wordsworth in *The Prelude* when he is hiking on Mont Blanc, but this has more interior, biographical significance than in *Briggflatts*. The narrator and his companion 'clomb with eagerness, till anxious fears / Intruded, for we failed to overtake / Our comrades gone before … every moment added doubt to doubt' (1979 [1850]: 217; part VI, lines 575–78). This becomes a scaling of the ego as much as an ascent of the mountain, and aspires to keep climbing even at the very moment it recognizes its failure: 'still we had hopes that pointed to the clouds' (ibid.: line 587). These ambitions are figured by Alexander in *Briggflatts*. He serves as a mythic symbol for, rather than a psychological dissection of, arrogance, in contrast to his more cautious soldiers. The emphasis in his ascent, and in the slowworm's song, is the exploration of the environment as much as the self.

This episode prepares the way for the dispersal of ego and contextualizing of human time in the subsequent movements of the poem, which work themselves out in the discrepancy between human ego and natural agency. Conniff remarks that, as *Briggflatts* progresses,

> The seasons of the year no longer have an obvious parallel in the seasons of the poet's life. He is caught, suddenly, in an anti-Romantic schism: he can no longer

assume a fundamental sympathy between his emotions and the natural world, a sympathy that would have allowed him, in effect, to subordinate a world of "objects" to his own subjective experiences. (1988: 175)

That is to say, the phenomenal world in *Briggflatts* resists the imposition of an order in the form of traditionally conceived natural seasonal cycles, much as anthropogenic climate change exposes human inability to master or engineer atmospheric processes. It expresses an awareness that human lives are both subject to natural forces and outlived by them. This serves as a context for the unseasonal, entropic motifs that occur in each of the ostensibly seasonal movements of the poem.

Without the poet's supervening lyric ego, Conniff contends that *Briggflatts's* 'natural forces are benign, even though they have been "let loose," as far as possible, from the narrator's controlling mind' (1988: 184). How far *is* 'as far as possible'? After all, Conniff treats Bunting's resistance to Romantic tendencies here as a conscious poetic strategy, a pose in which the poet only *seems* to relinquish control rather than actually giving it up. Nevertheless, even though Bunting tacitly draws out an increasing disorder in nonhuman nature, in doing so he counterpoints traditional ideas of natural harmony, because the emphasis of *Briggflatts* is on 'chance events' (Conniff 1988: 184). The poet resists the identification of individual life with natural, seasonal cycles. Once Bunting sees its fallacy, the seasonal cycle of *Briggflatts* dissolves with the final movement of the poem, which, as we have seen, begins with unseasonal melt: 'Drip – icicle's gone' (2016 [1965]: 58; part V, line 1).

Climate change gives unanticipated material reference to Bunting's disordered seasons, showing us that what we regarded as pastoral timelessness is in fact a product of our current interglacial episode. Bunting provides a salutary reminder of our earthly bearings – assuming we are prepared to accommodate rather than neglect or manage the climate in which we are enmeshed, with its own inexpressible, material agency. This agency is still more clearly seen in the work of Bunting's contemporary, David Jones, whose work I discuss in the next chapter.

David Jones's *Anathemata* and the Gratuitous Environment

Given that no feature of our biological or cultural existence can be isolated from climate change, no text can be ringfenced from being read in relation to it. Climate change criticism must therefore come to bear even where climate change is not the matter of the poem; climate cannot be bracketed off into the genre of the 'environmental', as it is in political and news media agendas, so neither should the scope of its relevance to literature. I have aimed to demonstrate that poetry pre-dating popular understanding of the phenomena of climate change recognizes (whether implicitly or explicitly) the contingency of human existence in the terrestrial environment, and that Modernist poets challenge defined, formal boundaries between the cultural and natural as well as the presumption that we can master climatic forces. While Modernist poetry reveals that we have always affected and been affected by the agency of nonhuman phenomena, this relation is revealed again, materially, in the emergence of contemporary climate change. Both text and climate expose the entanglement of agency in making the world – and indeed the difficulty of distinguishing between the two – and this can also be demonstrated through a critique of a Modernist work that *is* explicitly occupied with the human–climate relation: David Jones's *The Anathemata*. The strategies employed in this work represent a creative engagement with the intellectual challenges I have thus far outlined.

This work[1] addresses the development of earth's prehistoric environment into conditions suitable for humanity, describing the genesis of terrestrial landscapes and cultures after the end of the preceding glacial. *The Anathemata* is concerned with history's development from primeval origins, in particular in its first section, 'Rite and Fore-Time', and implicates environmental factors in the appearance of civilization. It traces the emergence of humanity in terrestrial environments and civilization's spread across the sea, within a sequence of oblique narratives – in Jones's subtitle, 'fragments of an attempted writing' (2010 [1952]: n.p. [3]) – where the motifs of Christ's incarnation and passion resound both backwards and forwards in time in accordance with Jones's Roman Catholic beliefs. In this chapter, I will concentrate on 'Rite and Fore-Time', which describes the end of the last ice age and the evolution of human culture and art, drawing in subsequent sections of *The Anathemata* where relevant. Because of its subject matter, the scope of Jones's work necessarily responds to the demands that climate change makes on the imagination and on literary form. I will show how his poetics enables expression of the agency of natural phenomena in human development, with my reading redirecting the focus of the work from Christ to climate.

Jones's process is to find images that resonate beyond specific, historic cultures, accumulating these motifs into a fractal work where individual vignettes and episodes present scale versions of a master narrative, making full use of his text's scope as 'open in form' but 'formally whole' in Thomas Dilworth's words (2008: 118). *The Anathemata* incorporates a diversity of forms and modes: the opening spread of the work proper (Jones 2010 [1952]: 48–9), for instance, presents the reader with inscription, prose and verse, all rife with quotation and parenthesis, reflecting the work's deliberately fragmentary quality. Interpretative direction is provided by Jones's preface and footnotes, some of the latter proving so extensive that they require full pages. This multiplicity of signification and the acknowledgement that *The Anathemata* is an 'attempted writing' together represent an imaginative engagement with prehistoric climate change rather than striving for an objective account.

Jones's array of techniques give considerable scope to the work. *The Anathemata* is able, for example, to encompass durations of time difficult to envisage on human scales, most pertinently prehistoric climatic change and the transition from the Pleistocene epoch to the Holocene. My attempted reading, as distinct from Jones's attempted writing, extends the implications of his account, conceptually and temporally, into the Anthropocene, the provocative and as yet unratified name coined for our present epoch by geologists who maintain that human presence on earth constitutes a geological influence in its own right.[2] Because Jones's poetics operates at a broader scale than individual human lives, indeed, than the lives of particular civilizations, he is able in Henry Summerfield's words 'to write lyrically of geological change and archaeological findings' (1979: 19). This capability can lend itself to a poetics of the Anthropocene, because it can assist us in making the difficult intellectual transition between our lives as individuals and the geological epoch we seem collectively to have instigated.

Jones's handling of our human narrative differs from more straightforward prose accounts of prehistory in a way that demonstrates what is possible in poetic, rather than scientific, engagements with climate change, complementing my analysis of Stevens's poetry as climate model in Chapter 3. The phenomena that civilization excludes in devising such science-based, progressive narratives are collected by Jones under the name of 'anathemata'; these exclusions can be either positive, in the sense of being venerated, or negative, in the sense of being overlooked or ignored. For Jones, the scope of the literary work is such that it can engage with these 'anathemata' in a way that science, constitutively discriminating, does not, and this chapter will show the value of such engagements to the writing of climate change.

My ongoing argument remains that the tensions in human relations with the environment that are explored in Modernist poetry resonate so strongly today because, exacerbated by another sixty or more years of human civilization, these tendencies put even greater strain on our attempt to understand or manage the world. To read works such as *The Anathemata* is thus to find a way of articulating those tensions, and to begin to trace their exponential development as the Anthropocene unfolds.

Poetry versus progressivism

Jones invites comparisons between *The Anathemata* and more linear accounts of human history in his citation, among many other sources, of *The Age of the Gods* by his contemporary, the historian Christopher Dawson. Dawson purports to offer an anthropological overview of the emergence and development of humanity and of civilization, promising 'to undertake some general synthesis of the new knowledge of man's past that we have acquired', thanks to which 'a general vision of the whole past of our civilisation has become possible' (1933 [1928]: xii) – though his endeavour has, like Jones's work, a Roman Catholic trajectory. To triangulate my reading of Jones and Dawson, I shall also refer to a more recent account covering roughly the same period as *The Age of the Gods*, Jared Diamond's *Guns, Germs and Steel*, which takes a scientist's rather than an historian's perspective. Diamond considers the terrestrial environment's influence on human development, surveying 'the 13,000 years since the end of the last Ice Age,' when 'some parts of the world developed literate industrial societies with metal tools' (1998: 13).

Both Dawson and Diamond are concerned with producing accounts that advance their cases through argument rather than literary imagination. As analytical narratives, they require themselves to make causal connections between data, but thereby tend to portray cultural development as a mechanistic sequence, one in which geology or archaeology is used to show how the course of events has been determined. Both tacitly recognize this possibility in moving to disavow it, however. Dawson, for instance, insists, 'Not that man is merely plastic under the influence of his material environment. He moulds it, as well as being moulded by it' (1933 [1928]: xiii). Diamond goes further: 'the notion that environmental geography and biogeography influenced societal development ... is considered wrong or simplistic, or it is caricatured as environmental determinism and dismissed ... Yet geography obviously has *some* effect on history; the open question concerns how much effect, and whether geography can account for history's broad pattern' (1998: 25–6; author's italics). Diamond's formulation asks how nonhuman factors and processes ('geography') have influenced culture ('history'). This is a useful restatement of the question that has occupied me throughout this book, given climate change's challenge to the notion that either nature or culture, geography or history, is a distinct or primary determinant of the world.

To see quite how accommodating *The Anathemata*'s scope is to these interweaving agencies, we can make a useful comparison with the modes of both Dawson's and Diamond's accounts. Both the latter authors seek to employ scientific discourse rather than overt theological signification to structure their narratives; but in so doing, both also take a more teleological approach. Instead of making Christ central to history, as Jones does in *The Anathemata*, Dawson and Diamond alike position humanity as the ultimate end of millennia of progress. Although Dawson maintains that 'progress is not ... a continuous or uniform movement, common to the whole human race' (1933 [1928]: xvi), his subordinate clause does not so much problematize his understanding of the concept as make it the preserve of particular peoples. He goes on to reflect that 'Progress is an abstract idea derived from a simplification of the multiple and heterogeneous changes through which the historic societies have passed' (ibid.), implying that it is a helpful if reductive organization of events, but again limiting it to those societies privileged enough to count as 'historic'. This is most evident in the linear metaphor he uses in his discussion of Neanderthals, whom he considers 'an over-specialised by-product, a side path or blind alley on the road of human development' (ibid.: 10).[3] By contrast:

> it was probably only after the expulsion of man from the Paradise of the Tertiary World, with its mild climatic conditions and its abundance of animal and vegetable life, that he made those great primitive discoveries of the use of clothing, of weapons, and above all of fire, which rendered him independent of the changes of climate and prepared the way for his subsequent conquest of Nature. (Ibid.: 6)

In these remarks, Dawson conflates Biblical and scientific accounts of history, dating 'Paradise' to the period from around 65 million to 2.6 million years ago,[4] to read the emergence of humanity as a gradual triumph over the environment. Note though that fire is brandished by Dawson as the most significant technology in our triumph over the climate, whereas contemporary science recognizes combustion as one of the major factors in our (continued) implication with it.

Like Dawson, Diamond is conscious that he may be read as endorsing a progressive account of history, and he tries to disclaim an ideological inflection to his subject matter by saying 'We tend to seek easy, single-factor explanations of success. For most important things, though, success actually requires avoiding many separate possible causes of failure' (1998: 157). Yet even here, talk of 'failure' suggests the grand narrative is one based on the accomplishment (or otherwise) of defined goals along what Dawson calls the 'road of human development'. Diamond makes repeated use of the terminology of failure and development in describing historical change. He begins his fifth chapter, 'History's Haves and Have-Nots', for example, by declaring that 'what cries out for explanation is the failure of food production to appear, until modern times, in some ecologically very suitable areas that are among the world's richest centers of agriculture and herding today' (ibid.: 93). Likewise, his account of the 'problems [that] *delayed* the domestication of apples, pears, plums, and cherries until around classical times' (ibid.: 125; my italics) seems to depend on the notion that history runs, or should have run, according to a schedule; this assumes the material conditions of geography should have a determining role in history, inasmuch as they ought in Diamond's view to have led to the development of certain practices. But this is a retrospective imposition. Diamond takes the data of Western history and makes them the yardstick by which he judges other cultures, underscoring an ideological leaning towards European ideas of progress.[5] Hence, he reflects on 'some puzzling non-inventions in the Americas' (ibid.: 370), faulting the pre-Columbian peoples for not achieving what their European counterparts had managed. He is even more explicit in lauding contemporary neo-liberalism when he describes, in contrast to China, the 'factors behind Europe's rise: its development of a merchant class, capitalism, and patent production for inventions, its *failure to develop* absolute despots and crushing taxation' (ibid.: 410; my italics). The prose modes of both Dawson and Diamond privilege a particular vector of human development towards a contemporary pinnacle, assuming the gradual transition of terrestrial influence from natural to human agency – from geography to history – as civilization progresses.

The Anathemata might also be read as moving eschatologically, towards an historical apogee, although it is divinely rather than humanly directed. Jones reads environmental factors as preparatory for the appearance of humanity, of civilization and of its saviour on the planet. Prehistoric changes in the terrestrial environment do literal groundwork for civilization, as in the following passage where Jones envisages the rise of the Welsh landscape and people prior to glacial melt:

> Before the Irish sea-borne sheet lay tattered on the gestatorial
> couch of Camber the eponym
> > > lifted to every extremity of the sky
> by pre-Cambrian oros-heavers
> > > for him to dream
> the Combroges' epode.
>
> > > > > > > (Jones 2010 [1952]: 67)

As Summerfield indicates, an 'eponym' is 'a person from whose name the name of a nation is supposedly derived. Cambria = Wales (Med[ieval] Lat[in])' (1979: 50); this identifies the individual with the land, privileging the former. In *The Anathemata*, the landscape serves to elevate the figure for whom it will be named, to let him compose the 'epode' that will unify his people, who are in turn associated with the land. The elevation is both physical and ritualistic, given the use of the word 'gestatorial', glossed by René Hague as 'originally merely a *sella gestatoria*, or sedan chair, but borrowing a grander sense from the papal *sedia gestatoria*' – the throne in which popes were borne on ceremonial occasions (1977: 60; author's italics). The movement of ground free of the zone of glacial melt enables the ascent of the civilization that develops on it.

Despite the orientation of this passage towards human existence, we can nevertheless distinguish the way that Jones refers to embodied experiences, such as being borne aloft in a papal chair, from Dawson and Diamond's more technical overview of human evolution and cultural progress. Moreover, it is in this passage the environment that uplifts humanity, rather than human will itself, giving it a – literally – elevated foundation on which to begin its literary endeavour, which reflexively celebrates the land. There are other instances in *The Anathemata* where a similar embodiment, rather than portraying the environment in terms of its human utility, actually enables nonhuman forces to be made legible, such as the beautiful image of the mythical figure of Lamia, 'Her loosed hair … marking the grain of the gale' (Jones 2010 [1952]: 107). More broadly, in the words of N. K. Sandars, 'Jones takes this great heap of the past and tells it not as history, but as something we have experienced in our own flesh' (1976: 53).

Significantly, Sandars differentiates Jones's practice from that of scientific discourse. 'This new, this larger and infinitely more complicated world', discovered by science, 'is intellectually known, but hardly yet *felt* at all'; therefore, 'just as soon as scientific knowledge has apprehended new territory it is proper that the poets should appropriate it' (ibid.: 51; author's italics); as in Stevens, the literary serves as a mode that can complement our technical models of the climate.

Though it has the capacity to inspire, or be inspired, by wonder, science necessarily constitutes a separation between the discovery of data and the wonder of that discovery, as Jones suggested in a letter of 27 March 1943 to Jim Ede: 'Incredibly "romantic" these exact scientific things are, and the more factual so much the more moving' (1980 [1943]: 122). Jones's technique in *The Anathemata* is not scientifically objective but a self-acknowledged fiction, making phenomena accessible to human imagination through the evocation of the senses while also emphasizing our coexistence with those phenomena. 'The new science' of the early twentieth century 'required and provided the pressure for the evolution of a new epistemology which could relate the abstract logic of the mathematical relations to the appearance of particulars, sense-data, in the world', Patricia Waugh explains (2001: 8). Jones responds to scientific discovery by positioning it within this understanding. His general note to *The Anathemata*'s opening section, 'Rite and Fore-Time' explains that 'The findings of the physical sciences are necessarily mutable ... But the poet, of whatever century, is concerned only with how he can use a current notion to express a permanent mythus' (Jones 2010 [1952]: 82). Dawson and Diamond both synthesize 'the findings of the physical sciences' but are much less explicit about the 'permanent mythus' of progress with which they frame those findings. Jones on the other hand is open about 'the embodiment and expression of ... that cultural complex' (ibid.: 19), which is catholic in both its denominational and broader senses.

The Anathemata's pattern is not limited to the trajectory of humanity's emergence from prehistory, though, and unlike Dawson and Diamond his account also speculates about what may come after civilization; that is to say, humanity is not the end point of his work. The cycles of glacial and interglacial that Jones charts through 'Rite and Fore-Time' therefore also extend into the future, beyond the human moment of the present interglacial. Though he charts the emergence of Troy as a prototypical city, precedent for the rest of civilization, he nonetheless casts his mind forward and asks whether it could end 'under, sheet-darkt Hellespont' (ibid.: 57), lines glossed by Summerfield as asking, 'Will glaciers one day cover Greece and the Aegean?' (1979: 41). As with Bunting's adoption of Chōmei or Eliot's litany of cities in *The Waste Land*, Jones eschews a progressive narrative of human development, instead positioning his work in the vicissitudes of climate, with the toppling of an emblematic civilization recurring across history. In imagining an iced-over future in the Mediterranean, he implicitly acknowledges the permanence of our geological, climatological and evolutionary contingency rather than celebrating a civilization independent of its environment as Dawson and Diamond do.[6]

Although both science and poetry attend to the same phenomena, then, they treat them in very different ways. Richard Kerridge usefully directs our consideration of the distinctive functions of scientific and literary approaches that are in play here. He writes that 'The environmental crisis is only identifiable by means of expert interpretation of immensely complex, *constantly changing* data, and by the use of computer modelling and specialized techniques of statistical analysis' (2013: 349; my italics). He strictly demarcates the remits of science and of culture, in the specific form of literary critique, by declaring that 'the scientific data and the interpretation of those data are fiercely contested in ways that only experts can evaluate', so 'part of the business of ecocriticism is to define how [the] taking-on-trust [of scientific findings] can be done scrupulously' (ibid.). This is not to define separate areas of interest for science and criticism, but to describe their differing responsibilities to common interests.

Kerridge therefore provides a critical response to a creative problem that Jones identifies in the preface to *The Anathemata*: the 'tempo of change … in the physical sciences makes schemes and data out-moded and irrelevant overnight [and] presents peculiar and phenomenal difficulties in the making of works' (Jones 2010 [1952]: 15). Jones's own response to the handling of data in *The Anathemata* is to provide an interpretative framework in which to organize and evaluate the science of his day.[7] The poet's juxtaposition of scientific data with his Catholic 'mythus' shows that the two serve different functions, which we can characterize as discovery and revelation respectively; not unlike the distinction between reasoned processes and sudden encounters in the last part of Stevens's 'Notes Toward a Supreme Fiction'.

One consequence of making this distinction is that we are not then required to have faith – indeed, 'believe' – in scientific findings in the way we might in religion or mythical narrative, because such findings are necessarily mutable. It is the cumulative weight of these, rather than individual data, that suggest a paradigm, but that itself is not a matter of belief as such. Our beliefs are not susceptible to evidenced argument in the same manner as our interpretation of the world is, though the attempt to co-opt climate change discourse to that of belief or denial is, as Alaimo points out (2010: 16), to try to conflate the two categories. While there are ample data to show that the climate is changing, and that human beings are largely responsible for this,[8] vested interests still promulgate denial; likewise, abundant evidence exists for the theory of evolution by natural selection, yet this has not prevented the emergence of so-called Creation Science. What are cast as equally valid beliefs in each case actually fall on either side of the fault line between science and religion: the first is an attempt to understand the mechanisms by which the cosmos operates, the second is an attempt to satisfy humanity's desire to feel at home in that cosmos. When we elevate the paradigm of science to the status of world-view or 'permanent mythus', as in scientism, rather than regard it as a means of critically investigating phenomena, we make a mythology of scientific practice itself, 'when it serves as revealed truth in which we need only believe without question' (Botkin 2012: xvi). Yet both Dawson and Diamond adopt scientific discourse precisely to normatize the myth of progress with which they inflect their findings.

In contrast, Jones's poetics formally enacts the recognition that there are agencies that exceed our understanding, and that our interaction with the world represents a transformation of its data to bring them within our comprehension. This is evident from *The Anathemata*'s opening lines – 'We already and first of all discern him making this thing other. His groping syntax, if we attend, already shapes' (Jones 2010 [1952]: 49). Although, as Summerfield explains, the start of the work concerns 'A very early prehistoric man performing a religious rite [who] already sets some object apart as a sign and thereby prefigures a Roman Catholic priest celebrating Mass' (1979: 41), the terms in which Jones describes him are broad enough to suggest the figure of the artist–craftsman who will recur through the work (not unlike *Briggflatts*' stonemason). The 'groping' opening lines of the work situate it at the point of relation between the human and the material – 'this thing' – the attempt to make the unsayable sayable.

The acknowledgement of language's role in transforming our experience of the world is most clear when Jones sets the scene for humanity's evolution. He responds imaginatively to a prehistoric cycle of glacials and interglacials, with the shifting landscape they create: 'where the world's a stage / for transformed scenes / with metamorphosed properties' (Jones 2010 [1952]: 62). These lines are as aware of their own artistry as Shakespeare's from *As You Like It* to which they allude (2.7.140), and Jones's performance here is double. First, his lines mount the scenery in which every subsequent performance of the Catholic rite of mass will be rehearsed. He develops this metaphor of environment as performance space in the lines that follow, which speculate on the moment that God distinguished humanity from the rest of the animal kingdom: 'from what floriate green-room, the Master of Harlequinade, … called us from our co-laterals out …?' (Jones 2010 [1952]: 63). In this image, God performs the role of an impresario, directing humanity's entrance on to the stage of the earth; Jones then plays with the concept of the 'green room' as a space for theatrical preparation, by opening it into natural space that is 'floriate', but is nevertheless 'room', functional space, in which humanity prepares to make its entrance upon the stage of history.

Second, Jones's projection into the past is a staging of one among any number of dramatic possibilities presented by the data of the fossil and archaeological record. A litany of questions precedes the invocation of a primeval theatre (ibid.: 61f.), and these emphasize the writer and reader's mutual, ritual imagination of prehistory, marking itself as hypothetical, or fictional in the sense of Stevens's projections into nature. The questions are quasi-scientific not in purporting to seek objective answers but by engaging in open inquiry. Jones's metaphor of changing environmental scenery anticipates the kind of staging considered by Ulrich Beck, where hypothetical scenarios of environmental collapse are drawn into the present in order to prevent their occurrence: 'only by imagining and staging world risk does the future catastrophe become present – often with the goal of averting it by influencing present decisions … a prime example being the debate on climate change which is supposed to prevent climate change' (2009: 10). Beck continues that 'the staging of global risk sets in train a social production and construction of reality. With this, risk becomes the cause and medium of social transformation' (ibid.: 16). Jones offers a model for attempting this in literature, in a performative rather than declarative mode, as well as instructing us in the art of dramatizing mythic premises for the present, but without having to suppose them to be objectively accurate in the manner that Dawson or Diamond would require them to be. Jones weaves a narrative into the data of the present and the past, and that in turn conditions his climate-contingent vision of the future.

Despite the seeming authority of scientific accounts of the world, we still, as a culture, have a need for mythical understandings of it such as Jones's. Botkin attests that, 'We need to see mythology—in the sense of a story about how the world came about and how it works—as still a necessary part of human existence' (2012: xvi). He suggests that science is a complement of and not an alternative to our mythologizing because 'even today, in this age when we seem to have persuaded ourselves that we have risen above mythology, most environmental policies, laws, and ideologies are consistent with (to say the least) and arguably a restatement of the beliefs about nature in th[e] Judeo-Christian tradition' (ibid.: xvii).[9] While Jones follows Dawson in explicitly framing prehistoric climate change with Biblical motifs, albeit without the latter's more evidently humanist teleology, Diamond adopts no ostensible mythus but nevertheless still inherits Dawson's progressive reading of evolution. Where Jones self-consciously uses narrative to enable us to understand climatic change, Diamond tries to sell us his narrative through a purportedly objective scientific account. But it does science a disservice if we regard it, as Diamond does, as sufficient to fulfil our need for mythus; Botkin emphasizes that our engagements with the world persist in having a sacramental quality, even when devoid of an explicitly theological context.

As a result, while it strives to remain objective, scientific practice is always hybridized by the account we would make of it. Hence, Szerszynski maintains, while 'it may seem scandalous to divert attention away from the task of a causal analysis of climate change, and instead to try to understand it in terms of semiosis and meaning ... it is the dominant technological framing of climate change that ultimately constitutes a more radical evasion of responsibility' (2010: 22). This is, he argues, because rational, instrumental analysis of climate presupposes that it will respond in predictable or manageable ways to our intervention. He continues, 'standardized forms of measurement, and ... conventional practices of aggregation and modelling' function by 'bringing the weather indoors', and this 'tempts us to imagine that we can discern a "divine writing" in nature' (ibid.: 22–3). We become peculiarly susceptible to the illusion of mastery over an objective reality once we have abandoned the notion of God, because by placing faith in scientific accounts rather than valuing them for what they are – methodically devised and rigorously tested propositions about phenomena – we make them into sacraments.

We thus overdetermine our interpretations of climate by investing humanity's position with an unacknowledged divinity. It is this misreading of our relation to the science that can lead us to a 'confident belief in the human ability to control Nature[, which] is a dominant, if often subliminal, attribute of the international diplomacy that engages climate change', as Hulme puts it, relating technocratic environmental governance to the myth of Babel (2009: 351). Unlike such scientism, which supposes to divest itself of mythology and ideology, Jones's poetic engagement with the emergence of civilization is explicit about its mythical, sacramental status, expressing the impossibility of sharply distinguishing human activity and thought in both the physical and cultural construction of nature.

The fractal form

Szerszynski argues that, '*Writing itself* (as the condition of im/possibility of meaning) is always aberrant, and reading the climate is thus always already subjected to the vagaries and aporias of writing' (2010: 22; author's italics). Because *The Anathemata* is itself 'produced from a simple formula (paradox, duality, aporia)', that results in 'a highly complex organism which repeats itself through each strata of its form' (Stanbridge 2011: 294), Jones's poetics works from its own aberrant quality. The poet uses this to trace the climatic contingencies with which *The Anathemata* engages, rather than make what Stevens calls a 'rational distortion' of the world in the way that Dawson and Diamond might be said to. Where purportedly scientific discourses such as theirs attempt to render the data of history into cause-and-effect progression, *The Anathemata* is in contrast all too aware of the multiple and various potential inflections and directions that narrative might take. Formally intrinsic to the work is thus an opening out from linear narrative into resonance with an environing culture.

This is especially evident in part V of *The Anathemata*, 'The Lady of the Pool'; this opens with a mariner's arrival in a mythologized London, before it recounts the song of a lavender seller whom that sailor may have heard in the city, nested within which is in turn the florist's vernacular inventory of local churches (Jones 2010 [1952]: 124–7); the different speakers then resume their respective relations as the section continues. The technique creates room in narrative for considering relations between several scales, while Jones's notes also prompt us to think about the text's various significances. As a way of reading the world, this can also enable us to account for the multiple relations and consequences that our engagement with the environment entail, something that becomes a formal issue in the writing of climate change. Adam Trexler writes, for instance, that 'it would seem that a global process like climate change could be described anywhere and that there would be hundreds of strategies an author might adopt' (2015: 78). We can see, too, that Elizabeth Kolbert has consciously to limit the scope of her account, *Field Notes from a Catastrophe*, claiming that she 'could have gone to hundreds if not thousands of other places ... to document' the effects of climate change (2007: 2). As a result, her narrative is structured around her experiences and research as a journalist, inviting us to read climate change in the same way we read journalistic accounts of other subjects.[10]

In contrast, a form that accommodates and even depends on the exigencies with which climate change presents us, as that of *The Anathemata* does, offers a more accurate reflection of our contemporary risk society, where causes accumulate over time and space into emergent and unexpected effects. Structurally, poetic modes are capable of enacting the unexpected and multiple phenomena of climate change and their connections. Poetry that encourages us to think about multiple associations fosters a sense that each of our engagements with the proximate, experiential world entails a relation at other scales, too – an awareness, say, that by eating a peach we might disturb the universe – which would read as a paradox if we were to try expressing it in realistic or rationalist modes. Timothy Clark is right to observe that climate change requires 'language [that] would have seemed surreal or absurd to an earlier generation', language that 'enacts a bizarre derangement of scales, collapsing the trivial and the catastrophic into each other' (2011: 136), at odds with the decorum of nonfictional prose narratives.

To assess the implications of this form, we can consider Stanbridge's 'proposed analogy for *The Anathemata* – the fractal' (2011: 392). Among the 'main characteristics of fractals', Stanbridge notes, are that 'they are self-generative from within themselves as a result of their iterativity' and 'self-similar in a hierarchized set of scalings' (ibid.: 398). This quality of self-generating resemblance across different scales means that *The Anathemata* presents the possibility of reading various motifs or themes as being dominant, recurring as they do throughout the work. Dilworth for instance regards the rite of mass as being formally central to the work, evident in the Lady of the Pool's 'lyrical celebration of the redemptive acts of Jesus, which the Eucharist sacramentally makes present' (Dilworth 2008: 177; citing Jones 2010 [1952]: 156–7). Like a fractally enlarged version of the pattern in 'The Lady of the Pool', Dilworth represents the pattern of *The Anathemata* visually as a nested structure organized around a core, 'as a recession of eight parentheses bracketing the lyric centre:

((((((((O))))))))' . (Dilworth 2008: 177)

However, the concept of iterative, fractal form also enables the work to be read eccentrically, even ecocentrically; in this case, the changing environment of 'Rite and Fore-Time' could be regarded as *The Anathemata*'s thematic as well as formal starting point. Such a reading, in contrast to Dilworth's and Stanbridge's, could be represented thus:

O))))))))))))))))).

This would make the work's own contingency, and that of civilization, more apparent by seeing the planetary 'O' with which *The Anathemata* begins ripple throughout the rest of the work. Such a reading does not reject Jones's patterning of the text, but regards that structure as open and responsive to its environment, as is *Briggflatts*.

Stanbridge uses the term 'fractal' to suggest Jones's compositional process is organic or phenomenal rather than planned on rational principles: 'Jones's text was generated by itself. Reading over his work, a fragment suggested a free-associational chain of other fragments' (Stanbridge 2011: 136). Indeed, Jones himself wrote to Harman Grisewood on 31 May 1938 that 'it seems to me that if you just talk about a lot of things as one thing follows on another, in the end you *may* have made a shape out of all of it. That is to say, the shape that all the mess makes in your mind' (Jones 1980: 86; author's italics). The result of this seeming 'mess' is that *The Anathemata* offers a new kind of spatial model that transcends several scales but finds similar patterns at those different scales. Timothy Morton also claims that: '"Text" is precisely the word for th[e] fractal weaving of boundaries that open onto the unbounded' (2010b: 2). *The Anathemata* is a site where these fractals interact at a scale visible to the reader, gesturing ever outwards at their repeated phenomenal formations, transgressing boundaries between history and geography, culture and nature.

To take up one instance of this fractal repetition through the work, we can consider some of the motifs by which it is evidenced. Jones asks in the preface, 'If the poet writes "wood" what are the chances that the Wood of the Cross will be evoked?' (Jones 2010 [1952]: 23). This complaint framed as question could be a problematic for the ecocritic, who will want to attest to the manifold biological and ecological associations of wood, rather than its significance in a particular religious tradition.[11] As his prefatory argument continues, though, it is clear that Jones is not so much channelling our response to language as (re)introducing another current in it. If readers do not pick up on the religious connotations of 'wood',

> that particular word could no longer be used with confidence to implement, to call up … the mythus of a particular culture … It would remain true even if we were of the opinion that it was high time that the word 'wood' should be disso-ciated from the mythus and concepts indicated. The arts abhor any loppings off of meanings or emptyings out, any lessening of the totality of connotation, any loss of recession and thickness through. (Ibid.: 23–4)

That is, the project of literature is for Jones a way of ensuring that language remains associative, retaining its ability to be iterative.

This is taken up in *The Anathemata*'s second section 'Middle-Sea and Lear-Sea', for example. On the voyage of a boat to Britain, as civilization and Christianity are spread through the world, Jones uses the terms 'mast-tree' and 'steer-tree' (ibid.: 102) to describe parts of the vessel. Summerfield says these terms stand, respectively, for 'the pole of the mast (perh[aps] the poet's coinage)' and 'the beam of the tiller (obs[cure])' (1979: 67). This neologism and archaism aim to evoke the wood of the Cross, as per Jones's preface, and the ship's timbers operate through the poem as a reminder of Christ's sacrifice, bearing humanity across the world's waters through time. The elision of 'wood' with 'tree', however, points up the biological constitution of those timbers, alluding to Christ's life-giving power and his vernal resurrection, just as the trees themselves come into leaf in the spring. What are ostensibly awkward or esoteric synonyms deployed for mast and tiller actually redirect attention from their instrumental function and remind us of their organic origins. Indeed, in *The Anathemata*'s seventh section, 'Mabinog's Liturgy', 'an old baulk' or timber beam is planted in the ground in place of a tree, and we are asked whether 'sawn-off timbers blossom ... Can mortised stakes bud?' (2010 [1952]: 190). Jones takes a view of the tree through time that resembles David Wood's understanding of the 'temporally extended persisting, growing tree' (2005: 152), and the poet's extension of the individual image through time into its own past and possible future fractally resembles the reappearance of wooden motifs across *The Anathemata*. Throughout the work, Jones seeks to revivify language through the resonance of the symbols he chooses.

Jones organically discovers his form through these motifs, as Stanbridge claims, rather than imposing it in advance. This kind of adaptive strategy has an analogue in the process of evolution with which *The Anathemata* is itself concerned; as Morton comments: 'Evolution shares pointlessness with art, which at bottom is vague and purposeless' (2010a: 44). Art in this reading is the expression of contingency, rather than an anthropocentric co-option of teleological narrative forms. So, while Jones sets his understanding of evolution in a Catholic context, he is not using it to justify a presumed human superiority. He remarked in a later letter to Grisewood (15–18 November 1970), 'the fond but sustaining belief of most men ... that notion of things "getting better" in every way [, was] borrowed from the evolutionary theory' (Jones 1980: 229–30). His use of the word 'fond' and the positioning of 'getting better' in quotation marks suggest that he doesn't consider we ought to read evolution this way, or at least accept that this belief *is* a belief rather than a scientific principle. *The Anathemata* shows, unlike Dawson, that 'The theory of evolution transcends attempts to turn it into a theological defense of the status quo' (Morton 2010a: 37).

Jones's concern to reinvigorate language runs counter to the persistence in modernity of what Robert Pogue Harrison calls our rational 'efforts to reduce the world to intelligibility through mathematics or history' (1993: 147), efforts that are evident in the work of both Dawson and Diamond. Rendering the world intelligible through instrumentalizing language is read by Harrison as symptomatic of the post-Enlightenment world, and is, for him, complicit in the process where forests are 'stripped of the symbolic density they may once have possessed', before 'an even more reified concept' came into play: 'the forest as a quantifiable volume of usable (or taxable) wood. The usefulness of the forest becomes measured in terms of a quantifiable mass' (ibid.: 121, 122). There is a common root to the thinning out of language and nature's reduction to its resource value: we neglect the resonances and histories of language to value its denotative qualities in the same way our use of so-called 'natural resources' neglects both their organic origins and a large, wasted portion of the subsequent output when they are combusted, an output that chiefly comes in the form of carbon dioxide. So, rather than imagining ever-outwards and thus recognizing that such emissions are necessarily associated with silviculture,[12] instrumental views of the world try to transact such effects into a non-existent externality. As Beck points out, 'the "side effects", which were wilfully ignored or were unknowable at the moment of decision, assume the guise of environmental crises that transcend the limits of space and time' (2009: 19).

In contrast to this reduction of significance, *The Anathemata* deploys specific references to give them a wider resonance, whether cultural, national or religious, and images exceed a single function, as in Jones's own example of 'wood'.[13] This corresponds to Peter Howarth's more general characterization of Modernist form in which 'ordinary material [is] given artistic charge by being poetically framed by structures in which no item or sound is ever subordinated into mere detail' (2012: 25). Jones's notes themselves provide such a structure, continually reinforcing the wider associations of particular instances. Yet, while much has been made critically of the status of the notes to *The Waste Land*, those in Jones's work are comparatively little discussed, even though they are considerably more substantial than Eliot's and run throughout *The Anathemata*. They represent another strategy in Jones's reassociation of connotation: rather than suppress irrational associations, he attends to the material agency of language. So, whereas Maud Ellmann characterizes *The Waste Land*'s notes as 'a kind of supplement or discharge of the text', representing the 'invasion' of the poem by its own 'disjecta' or waste matter (1987: 98), Jones's practice more deliberately constitutes an inclusion of the excluded in the work.

This is clear from his choice of title: 'I mean by my title as much as it can be made to mean ... the blessed things that have taken on what is cursed and the profane things that are somehow redeemed ... things, or some aspect of them, that partake of the extra-utile and the gratuitous; things that are the signs of something other' (Jones 2010 [1952]: 28–9). His concept of the 'extra-utile' here productively resembles Bonnie Costello's understanding of superfluity, which she reads as 'central to the principle of change in nature and culture' (1998: 569). Jones's transubstantiation of anathematized material into venerated text is then a celebration rather than a suppression of material agency, akin to the 'productivity [that] puts us in touch with the fluency of the universe' (Costello 1998: 569).

Thomas M. Lekan also argues that this quality is characteristic of fractal modes. He writes that 'as we contemplate the ecological and ethical ramifications of the Anthropocene, fractals do challenge any simplistic homogenizing or scalar reconciliation of the local and the global – and *point to an inexhaustibly exuberant nature* beyond doomsday narratives of finitude, revenge or atonement' (Lekan 2014: 195; my italics). Rather than have us rationalize or 'homogenize' the contradictory forces of the Anthropocene into a progressive model of history, in which events are oriented largely or wholly towards human development as they are in Dawson and Diamond, Jones's conception of the extra-utile, the superfluous, is replete with unrealized potential – as is the literary text itself. Indeed, Lekan's argument continues by proposing that fractal modes offer an 'effective index of ... post-Holocene critical practice, energizing an earth-wide network of connections amidst a fragmented, unequal, and exuberant world of difference' (2014: 197).

By opening continually out into signification, *The Anathemata* recognizes and enacts what Wood refers to as the 'end of externality':

> Now there is no outside, no space for expansion ... no slack, no 'out,' or 'away' as when we throw something 'out' or 'away' ... Yet so much of our making sense, let alone the intelligibility of our actions, still rests on being able to export, exclude, externalise what we do not want to consider. When that externality is no longer available, we are in trouble. (2005: 172–3)

Jones's poetics, like Bunting's, maintain an open-ended relation through the reader to the world of which they are both part. Considered in this context, where the extra-utile or superfluous represents a source of creativity, Jones's own notes then constitute not so much a closing down of the text but an obsessive recovery and recycling of associations that civilization has attempted to discard and disregard, restoring an excessive quality to his references. The notes are thus integral to Jones's process, rather than a retrospective manoeuvre as Eliot's notes to *The Waste Land* were.[14]

Recognizing the agency of these anathemata, Jones is therefore suspicious about instrumental views of the natural environment. For example, when fossil fuels were formed in the prehistoric past by 'the slow estuarine alchemies [that] had coal-blacked the green dryad-ways over the fire-clayed seat-earth along all the utile seams' (Jones 2010 [1952]: 72), that past is described in relation to its value for subsequent civilization, as with the earlier passage about the Combroges (ibid.: 67). However, Jones refers to the coal seams as 'utile', in contrast to the 'alchemical' processes that led to their formation. Civilization has neglected the gratuitous, excessive quality of the fossilization, and hence sidelines the emission of greenhouse gases that results from the coal's combustion, retrospectively inflecting Jones's ascription of 'utility' as a critique. These emissions are anathemata in the negative sense, a function of the positive, organic superfluity of the forests that become fossil fuels over geological time.

An identification between human and natural superfluity or creativity in *The Anathemata* becomes more problematic when Jones seems to limit creative activity to human beings: 'the extra-utile is *the* mark of man' (ibid.: 65 n.2; author's italics). This limitation is reinforced in his essay on 'The Utile', which defines the eponymous concept as 'the best word to cover the wholly functional works of nature, whether animalic or insentient (e.g. nest-building or mountain-building) and such works of man as tend to approximate to these processes of nature' (Jones 1959: 180–1). His understanding of the 'processes of nature' is thus difficult to align entirely satisfactorily with the idea of superfluous, excessive nature identified by Costello. Yet Jones's phrase 'wholly functional works of nature' implies a distinction with *non*functional works or processes. In a more nuanced elaboration of his definition, he writes: 'It is important to observe that the works of animals and of insentient creation, though wholly and inevitably "utile" in the fullest and best senses of that word, are impatient of being "utilitarian"' (ibid.: 181). The differentiation of 'utile' from 'utilitarian' in this context offers a distinction between instinctive creativity and informed intent. As such, the former has an excessive quality that the latter, more instrumentally, does not. What Costello refers to as 'the purging and renewing functions of superfluity' (1998: 572) are in this context the qualities of art and animal alike.

The associative Anthropocene

Jones's reaction to instrumental perceptions of the environment places him in a post-Romantic tradition that seeks to revivify rather than rationalize our experience of the world. Donald Worster places twentieth-century environment-alism alongside this history, which includes 'many biocentrists, Romantics, and arcadians' (1985: 333). According to Worster, these all responded to the way that, following the Enlightenment, 'Nature had been abruptly exiled by the scientific mechanists from the realms of value, ethics, and beauty' (ibid.: 318). Jones, too, participates in this project of restoring to the world the qualities of enchantment and signification by refusing to reduce terms to a single function – in his own words, to be 'not *realistic*' (Jones 1980: 80; author's italics).

The intensification and resonance of individual experience can be seen in the Lady of the Pool's relation of another sailor's voyage in part V of the work. She modulates between the seaman's technical reading of the weather as 'behaviours of water-spheres and atmospheres, as: incidence of tide and peculiar pressures of the upper air' and her own more evocative, sensual terms, 'Shifts of unshaping mist', 'muffle of grey fog' and 'Thicks of rain' (Jones 2010 [1952]: 139–40). Once again, scientific finding becomes reassociated with bodily sensation. Here, Jones continues to bring the weather into a personal relation that begins with the Lady's earlier lines: 'Come buy my sweet lavender / that bodes the fall-gale westerlies / and ice' (ibid.: 125). In that vernacular 'boding', the flower signifies weather to come and is not commodity alone,[15] as well as resonating with *The Anathemata*'s other references to vegetation. The work's capacity to 'convey' the reader 'imperceptibly from the shallows of our own experience of wind and weather onto quite different levels' (Sandars 1976: 67) is enabled because an individually textured voice such as the Lady's is also associative, becoming typical in the context of the work's global scope. It is not the identity of individual figures that unites, but the continuity and repetition of them as motifs. Similarly, Hague observes that 'the shipwright [of part IV, 'Redriff'], like the skipper in this and the preceding parts, is both individual ... and typical: thus he can serve to reflect the poet's shifting view-point' (1977: 148–9). Jones achieves this conjunction of character and context by repeated use of references that rehearse the liturgy, in fractally self-similar fashion, until the reader is inundated with them.

Such globalizing of individual significance is taken up by Morton in his notion of 'Ecology as text, text as ecology': he suggests that the fractal commonality of text and ecology entails an 'absence of a background' because the same principles generate both. On 'the globally warming Earth ... there is no longer any background ("environment", "weather", Nature and so on) against which human activity may differentiate itself' (Morton 2010b: 5). We have already seen how, in the 'Merlin-land' drawing, Jones establishes an equivalence between background and foreground, eschewing conventional perspective; likewise, *The Anathemata*'s own ecology of text also affirms a trans-historical order based around Catholic rite, in which the self is always situated in a wider pattern.

This density of association can be challenging, however,[16] and even Jones himself doubted he could successfully achieve what he intended. While conceiving *The Anathemata*, he wrote to Grisewood (14 February 1938): 'for the kind of writing I want to do you really do have to have so much *information* and know such a lot about *words* that I can't really believe I can do it except in a limited way' (Jones 1980: 83; author's italics). If this quality does mark one of the failures of Jones's writing, or our own response to it as readers, it still reminds us of what needs to be at stake in a poetics of climate change: the understanding and expression of inordinate, disordered global associations in every individual act. In this case, literary form may only be reflecting the potential for discomfiting disjunctions when immediate experience is juxtaposed with global significance. So, when Jones is dealing with the geological timescale of prehistoric climate change rather than bad weather in London, he attempts to manage these disjunctions by scaling processes to historic, human time to make them legible. This technique can be valuable to our comprehension of the telescoped timescales of anthropogenic climate change: whether this change is, or will continue to be, gradual or abrupt, it collapses industrial civilization's combustion, forest clearance and so on into material manifestation.[17] Each year of this in turn telescopes a greater order of prehistoric time into it, as Tim Flannery indicates: 'over each year of our industrial age, humans have required several centuries' worth of ancient sunlight to keep the economy going. The figure for 1997 – around 422 years of fossil sunlight – was typical' (2006: 69).

For Jones, poetry necessarily affirms the associations of material substance at any given point. Though he refers to both wood and coal in *The Anathemata*, he is at his most tellingly chemical when he considers water, and in particular 'whether the poet can and does so juxtapose and condition within a context the formula H_2O as to evoke … further, deeper, and more exciting significances *vis-à-vis* the sacrament of water, and also, for us islanders, whose history is so much of water, with other significances relative to that' (Jones 2010 [1952]: 16–17; author's italics). What a 'knowledge of the chemical components of this material' can evoke (ibid.: 17) would be an apt enough criterion for a poetics of climate change even if we continued to focus on water. It can be much greater if we substitute CO_2 into Jones's formulation, because then its significance extends to the whole planet's future rather than British history alone (as Jones means by reference to 'us islanders'). Our responsiveness to language's evocations can in turn help foster a greater environmental literacy, by encouraging us as readers to be aware of our position within material networks, and to gauge better the extent of our relations with them.

Another strategy for extending language's reach in *The Anathemata* is Jones's technique of shifting parts of speech from noun into verb, which transforms them from object or state into ongoing process. For example, he exploits a typical association between the female and the generative when he says of the emerging British landscape 'She must marl' and 'she must glen' (ibid.: 70). That 'she' may ambiguously refer back to the 'dim-eyed Clio' or 'naiad Sabrina' (ibid.: 68, 69), drawing together the emergence of human history with its physical geography, in the form of their associated muse and water-sprite respectively. On the other hand, Jones may be using 'she' to refer to a generic, generative female quality in the same way 'he' is the type for the male craftsman or priest; the poet uses pronouns to similar effect across the work to allude to more than one figure, in the same way as the Lady and the sailor are both individuated and typical.[18] The effect is to make instances of individual agency both cultural in their resonance and collective in their effect, responding to the difficulty of expanding lyric's intense individuality into cumulative, unintentional action and global-scale physical phenomena that the notion of the Anthropocene necessitates – humanity as geological force. With the ascription of subjecthood and agency to landscape in 'She must marl' and 'she must glen', then, Jones invokes an environment as process, 'the noun being used here, as so often, as a verb' (Hague 1977: 66). By not regarding landscape as given, Jones also satisfies one of Lawrence Buell's original criteria for an 'environmental text', exhibiting a '*sense of the environment as a process rather than as a constant or a given*' (Buell 1995: 8; author's italics). Similarly, in evolutionary rather than geological terms, reference to 'the mammal'd Pliocene' as a stage of evolutionary time (Jones 2010 [1952]: 74) makes the noun 'mammal' into a verbal participle, transforming taxonomic construct into an process contingent on a particular geological epoch.

These techniques expand the range of reference beyond human terms. Once again, though, Jones is able to scale up our thinking and make these vast timeframes in some way comprehensible in *The Anathemata*. He uses the seasons as a way of negotiating up from human calendars into geological time, as in the parenthetical passage that opens with reference to 'Great Summer' and 'Great Winter' (ibid.: 55–8). These are glossed by Jones as a 'Greek guess as to the cosmic rhythm ... largely verified by modern physical science' (ibid.: 54), also drawing on Dawson (1933 [1928]: 4). Within the first few pages of *The Anathemata*, the seasons are thus fractally scaled up into human epochs and ultimately, the 'cosmic rhythm'. As to the 'Great Winter', Jones comments that: 'I have no idea if at some remote geological time from now, there is any possibility of a similar glaciation ... I am merely employing such a possibility as a convenient allegory' (2010 [1952]: 58). This reading emphasizes the way Jones is able to consider civilization in a climate-contingent position by speculating about a further ice age, rather than imagining, as Dawson does, that civilization renders us 'independent of the changes of climate' (1933 [1928]: 6). The poet's imagination entertains the possibility of unexpected, phenomenal change rather than progress.

Human terms are further entangled in the geological in a subsequent verse-paragraph that again makes a speculative projection into the past, seeking the identity of the early human creator of the Willendorf Venus: 'whose man-hands god-handled the Willendorf stone / before they unbound the last glaciation / for the Uhland Father to be-ribbon *die blaue Donau*' (Jones 2010 [1952]: 59; author's italics). Charles Tomlinson comments that 'already, in the Venus master, we have the essentials of man-the-maker, in this pre-Teutonic world before the Uhland Father ... – himself a prefiguring of the Christian god – and the melting of the glaciers into the blue Danube' (1983: 12), a reading that distinguishes and juxtaposes human artifice and the melting of the glaciers. In Jones's formulation, however, the agency of the two parties is elided in 'they', which 'unbound the last glaciation'. The pronoun seems to refer anaphorically to the 'man-hands' that 'god-handled' the Willendorf Venus into being, entangling the emergence of human creation with environmental change. Dilworth reads the passage as 'the Sky Father melted Danube ice' (2008: 124); yet the question of exact responsibility for the changed climate remains unresolved due to the openness of Jones's poetics, which identifies the creative power of God with that of the first artists. Likening Jones's practice to 'what, in Biblical interpretation, is called typology' – whereby Old Testament figures were for instance presumed to foreshadow the narrative of the New – Tomlinson finds 'a whole structure of typologies' in the poet's work (1983: 11). Man the maker as a 'type' for God in *The Anathemata* emphasizes humanity's power to unmake the natural world as well as put it to productive use, and the material emergence of anthropogenic climate change subsequently certifies that resemblance by making a causal connection apparent. The emergence of humanity changes the planet on a divine scale, instigating the Anthropocene, but humanity still lacks a theology of itself that would enable it to comprehend this self-imposed role.

This proposed transition from the Holocene epoch to the Anthropocene begins with the movement from a sacramental to a more utile understanding of the world. Within the theological schema of *The Anathemata*, activities can be simultaneously sacramental and utile, which is to say, anathematic in the positive sense, or venerated. When glacial melt is figured by Cronos who 'breaks his ice like morsels, for the therapy and fertility of the land-masses' (Jones 2010 [1952]: 69), he foreshadows the rite of mass and also creates aquifers; while human agency, in the form of lighting 'aid-fires' in the cold, dark months of the year (ibid.: 221), is designed to 'help the sun survive the winter' as much as it is for generating warmth as Summerfield explains (1979: 134). Both actions have sacramental qualities, because they are not limited to their utile function but express a participation in a wider system. Once similar activities become regarded as solely utile, however, for the purposes of irrigation or power generation, we can industrialize them but neglect their anathematic qualities. Nevertheless, those qualities persist, but take a negative value. By ignoring their consequences, we then ironically reaffirm a cycle of positive feedback in which combustion intensifies the solar radiation received by the earth and the breaking of ice.

Similarly, the third-person possessive in the following lines depends on the shared purpose of man and God in the evangelical voyage: 'And now his celestial influence gains: / across the atmosphere / on the water-sphere' (Jones 2010 [1952]: 95). 'His' is used by Jones to suggest both the spread of Roman civilization and Christianity's influence around the globe. However, emptied of its theological significance, the possessive acquires a more utile, human reading. In that context, 'celestial' need have no more significance than 'atmospheric' would have, and retrospectively hints at the unintended influence of human activity on and in the sky. Jones writes of the fleet that 'the build of us / patterns dark the blueing waters' (ibid.) to figure its passage across the sea, and this resonates now as an image of 'built' civilization's projection of an environmental shadow, darkening the marine environment. Where *The Anathemata* aims for a wider, theological reading, it also enables future, secular readings where humanity occupies the space vacated by absent divinity. But narratives of progress such as Diamond's do not empower us to make this transition responsibly.

The discrepancy between sacramental and secular views of the world in Jones's work is picked up by Peter Chasseaud in his account of the poet's work for the Field Survey Company in the First World War. Chasseaud finds that the practice of mapping expresses the tension between multiple and single – anathematic and utile – significations in geographical terms:

> Jones was interested in the opposition between the Celtic view of the living land … and the Roman (i.e. modern) view of the land as an exploitable resource, to be measured, gridded, parcelled up, carved through with straight roads, and so on. He understands and accepts the utile technology … but is also disgusted by it. His sympathy is elsewhere, with the extra-utile, the sacramental, the mysterious. (1997: 30)

The poet's concern can be seen in his imagined account of German naval cartographers, and 'the greyed green wastes that / they strictly grid / quadrate and number on the sea-green *Quadratkarte*' (Jones 2010 [1952]: 115). Summerfield points out that Jones uses 'grid' here as a verb, meaning to 'cover with a grid', and likewise 'quadrate', or 'divide into squares', in the preparation of the *Quadratkarte*, a 'map marked with a square grid' (Summerfield 1979: 75). This verbalization emphasizes the process of mapping rather than the map itself, the contingent rather than definitive quantification of nature, akin to the process of continuously remodelling climates: the grid has to respond to the changeable quality of what is being mapped. The transition from prose to verse through these three lines of *The Anathemata*, where 'they strictly grid' establishes an iambic pattern for the hexameter line that follows, audibly demonstrates the play of cartography from irregularity into order as a process of artifice, attending to the human imposition of order on recalcitrant, excessive natural phenomena.

Contingent culture

As a work in language, in particular literary language, *The Anathemata* resists absolute closure or confirmation in its references to the divine, even if the author intends that the reader is able to recognize and understand these. Having created a context in which the ascent of humanity is contingent rather than entirely providential, *The Anathemata* is itself then subject to the same evolutionary forces. Comparing evolutionary processes' lack of direction and language's proleptic quality, Morton asserts that 'The reader is the future of the text … beyond and above the specific addressees of the specific message' (2010a: 80). Given that we are no longer obliged to read inherited religious stricture in the world as Jones does, we can still take from his articulation of humanity's relations with the environment a way of reading anthropogenic climate change. In his terms, it can be regarded as a product of the combined prodigality of human artifice – our negatively inflected anathemata – and natural process. The sympathetic, poetic text allows space to identify human anathemata and our complicity in environmental change, as well as culture's contingency in that environment. Jones's deployment of Modernist poetics in *The Anathemata* is testament to the scope this mode of writing affords to express the complex entanglement of human and natural agency in climate change.

The Poems of Our Climate Change

By refracting our understanding of contemporary climate change through the lens of Modernist aesthetics, we see more clearly that what we would conventionally distinguish as the 'human' or 'cultural' and the 'natural' are actually less well-defined, more entangled agencies that shape the planet. This condition exemplifies the concept of hybridity theorized by Latour because our environmental crisis is the product of a political categorization of Nature, and subsequently 'the environment', as such. Latour illustrates this with the example of newspaper discourse, where 'hybrid articles ... sketch out imbroglios of science, politics, economy, law, religion, technology, fiction', even when 'Headings like Economy, Politics, Science, Books, Culture, Religion and Local Events remain in place as if there were nothing odd going on' (1993: 2). If we consider the changing climate as something happening solely in a category called 'the environment', imagining that this exists around rather than entangling human culture, then we overlook the continual complexities of our relationship with climate. But any such attempt to pigeonhole climate change as an 'environmental' issue is impossible: the fact that the discourse of climate change still exists, is still necessary, three decades after it entered the public consciousness reflects the fact that the phenomena cannot be successfully categorized and legislated.

Rather than reinforcing rigid divisions between culture and environment, Modernist works demonstrate the problematic quality of these conceptions, as in Eliot's and Stevens's poetry, as well as the exchange between human and nonhuman forces, seen in the writing of Bunting and Jones. Their poetics can thus as readily if not more readily be read in response to the challenges presented by thinking contemporary climate change than can the representations of natural landscape or environmental crises to which earlier phases of ecocriticism urged attention. By refusing to be absolute or definitive, Modernist works are instead characterized by the precision with which they recognize and articulate uncertainty, lending themselves to continual re-examination and reinterpretation. They help us to identify that our characterizations of the world are contingent, and that climate change emerges from the assumption that nature is both static and infinitely able to accommodate civilization's externalities.

By explicitly advocating a Modernist poetics of climate change, I have assumed that the prevailing understanding of the phenomena in contemporary culture is largely reductive, operating from false premises about Nature. Botkin contends that 'our laws, policies, beliefs, and actions continue to be primarily based on nature as a still life. This is all the more ironic in a society immersed in movies, television, and computer games that are dynamic, and cell phones that can take moving pictures' (2012: 8). While I concur with Botkin that Nature is too often reified, it is not ironic that we separate Nature from culture so much as inevitable, because, given the pace of twenty-first-century lifestyles, we project a longed-for stillness and harmony on to the nonhuman world. When poetry today, such as Motion's 'The Sorcerer's Mirror', continues in this vein to foster a vision of Nature that is separate from humanity, it is understandably going to have problems in engaging with the multiple phenomena of climate change, because it retains, in an unexamined form, the vestiges of the outdated ideas I have critiqued. But this latter tradition still characterizes a fair proportion of actual climate change poetry that has thus far been written. Following a brief look at the different trends in such poetry, this chapter will go on to analyse in detail poems by Jorie Graham that exemplify the tendencies I have outlined, and reflect on her creative achievements and poetics in relation to those Modernist authors I have already discussed.

Warming to the theme

Poems begin to take up climate change as a distinctive topic in the late 1980s and early 1990s, in the wake of the first 'greenhouse summer' in 1988 – the year, when, according to Hulme, 'the idea of climate change penetrated more deeply into popular culture in the West' as a result of 'a convergence of events, politics, institutional innovations, and the intervention of prominent public and charismatic individuals' (2009: 63–4). The importance of climate change was confirmed as one of the key issues discussed at the United Nations Conference on Environment and Development – the so-called 'Earth Summit' – in Rio de Janeiro in June 1992.

As a result of this topicality, early poems about climate change included politically framed satires such as Les Murray's 'The Greenhouse Vanity' (first appearing in the *London Review of Books* in May 1989; revised for Murray's 2003 *New Collected Poems*; Murray 1989: 8; and Murray 2003: 332) and Simon Rae's 'One World Down the Drain' (collected in *Soft Targets: From the Weekend Guardian*, in 1991, and reprinted in the Bloodaxe anthology of ecopoems, *Earth Shattering*, in 2007; Rae 1991: 53; and Astley 2007: 196), but global warming also occupies individual poems of lyrical reflection, such as Steve Ellis's epistolary 'Son to a father, 21st century' and Lavinia Greenlaw's 'The Recital of Lost Cities' (Ellis 1993: 48; and Greenlaw 1993: 15). Fleur Adcock's 'The Greenhouse Effect' meanwhile mediates between lyric and topical response. The former quality is explicitly filtered through the Romantic tradition, as she reflects that the city of 'Wellington's gone Wordsworthian again' (Adcock 2000 [1991]: 204; Astley 2007: 198), adding that the earlier poet would 'have admired it – / admired but not approved, if he'd heard / about fossil fuels, and aerosols' (Adcock 2000 [1991]: 204); like Motion, she contrasts an Anglophone tradition of nature poetry with human impact on the environment. These poems also mark early appearances by the arctic imagery recycled in 'The Sorcerer's Mirror' along with rising sea levels and parched landscapes; constituted from these tropes, however, global warming does more to offer new objective correlatives (understood in Eliot's original sense) for poets to talk about conventional, human concerns, such as inheritance or political ineffectiveness, than it does to consider what is distinctive and disruptive about climate change. Without a longer perspective or a reconsideration of tradition, there is no discovery of anything about climate change per se; the climate may be changing but, fundamentally speaking, our outlook is not.

The persistence of climate change both as phenomena and, to an extent, as an agenda, has seen it merit continued consideration in subsequent decades, though individual lyrics have remained a common means of engagement. A 2008 piece in the poetry magazine *Magma* suggests why this remains a popular approach:

> we may be entering a period when, for the first time since the Middle Ages, the quality of everyone's life will worsen owing to climate change. The extent of this and its impact on how people think and feel are unknowable as yet, but the experience of condemning one's children and grandchildren to a life worse than one's own may be difficult to bear ... Against this grim prospect poetry may seem slight but in fact it is very powerful ... Poetry crystallises people's feelings about themselves and the world, and if it can show people how to feel in new ways in response to unprecedented changes in the world, it will help us to survive. (Smith 2008: 22)

There is still a predominantly individual focus in Laurie Smith's remarks, as he still awaits climate change's personal 'impact' rather than recognizing that his concern is already one such effect, but the scope of change he envisages shows that global warming is now more than simply a 'topic' for poetry, and must command a more considered response.

The movement into intergenerational time in Smith's remarks, already made retrospectively in Ellis's 'Son to a father, 21st century', is recapitulated in Simon Armitage's 'The Present' (2010) when the narrator meditates on an icicle found in an upland English landscape and considers what it means for his offspring. Its Romantic inheritance is confirmed in the poem's receipt of that year's Keats–Shelley prize, and though its scope is lyrically constrained, it retains a Keatsian negative capability by hinting at the future that this 'present' moment portends, rather than stressing a more hackneyed imagery of arctic collapse; as such it remains open to engagement with the problems I have identified throughout this book. Still more successful in effecting an open lyricism in response to climate change is a triptych of poems by the late Seamus Heaney – 'In Iowa', 'Höfn' and 'On the Spot' – from his collection *District and Circle* (2006: 52–4), which attest to the global extent of the phenomena by locating their narrator in a Midwestern cornfield, in flight over Greenland and in his own garden, refusing to limit themselves to visions of untouched Nature and, indeed, by taking a view from an aeroplane, tacitly acknowledging the implication of personal travel habits in the scene observed below. Though each of the poems then depends on the observing, experiencing 'I', their juxtaposition, without explicit connection, makes them a more associative encounter with climate change than the concentrating vision of 'The Sorcerer's Mirror', leaving the reader to connect them.

The potential that longer forms offer to engage with climate change has also been taken up in different modes, for example the postmodern collage of Peter Reading's collection *−273.15* (2005),[1] or the knowing, ironic elegies of D. A. Powell's *Chronic* (2009). Reading's collage technique and Powell's allusions to *The Waste Land* – 'nobody said the undertaker would come spanking new in a blinding heat / his crucible searing arctic glaciers [indeed: *summer surprised us*]' (Powell 2009: 43; author's parenthesis and italics) – point to the development of alternative, more self-reflexive modes to engage with the complexity of the phenomena and where to situate them in the literary tradition. Peter Reading is also ambivalent about the value of co-opting Romanticism in the encounter with climate change, recalling 'The school mag. *Juvenilis* piece, "Bonfire", a puerile Keatsesque thing, proved microcosmic after all' (2005: n.p. [21]).

The intensification of interest that these more recent poems represent corresponds with an increased attention being paid to climate change, beginning in the middle of the first decade of the twenty-first century, before the credit crunch and banking collapse. This concern was prompted by debate around the causes of Hurricane Katrina in August 2005, and is also likely to have been influenced by the publication of the Stern Review on the Economics of Climate Change in the UK (2007) and the Intergovernmental Panel on Climate Change's (IPCC) Fourth Assessment Report on Climate Change (2008), as well as the major films *The Day After Tomorrow* (2004) and *An Inconvenient Truth* (2006). The continuing Conferences of the Parties to the UN Framework Convention on Climate Change, such as COP15 in Copenhagen in 2009 and in particular COP21 in Paris in 2015, have kept climate change intermittently on the news agenda since then; and in the run-up to the latter, another facet of the poetry of climate change has become apparent, specifically its serving as an institutional response.

Among a range of public initiatives that involved commissioning and publication of new work were the Royal Society of Arts's Seven Dimensions of Climate Change, and Weatherfronts: Climate Change and the Stories We Tell, the latter being one of a number run by London-based literary organization the Free Word Centre. Similarly, just as Andrew Motion's 'The Sorcerer's

Mirror' was a commission serving a public function, a similar association hangs over his successor as the UK's poet laureate, Carol Ann Duffy, who gave her imprimatur to the *Guardian* newspaper's Keep it in the Ground campaign,[2] launched in spring 2015 and commemorated by the re-publication of her poem 'Parliament'.[3] Duffy then oversaw a series of twenty-one poems on climate change, 'Our melting, shifting, liquid world' that appeared on the *Guardian*'s website in the months leading up to COP21. Making available audio of the poems as read by well-known actors, with the option of viewing the text next to portraits of the performers, the project is mainstream in its aim, and the poems, by established contemporary poets, largely more conventional than experimental as a result.

The nature of these and other projects as commissions also remains problematic inasmuch as it confirms rather than challenges the constitution of climate change as a political issue on Latour's terms, and poets' or editors' attempts at persuasion or motivation are then stymied by being unlikely to reach those readers who are not already disposed to take an interest. If we engage with them, we may opt to feel worthy for having read poems about climate change – or, conversely, to ridicule those who imagine that poetry might be an effective way of overcoming it. If on the other hand we prefer not to engage with climate change, to deny human complicity in it or dispute its existence, we can just as easily choose not to read the anthology. Its function is framed and constrained by its sponsorship by an institution or media outlet, and by the project of anthologizing: the choice is more than likely to confirm our existing ideologies rather than challenging them. If that work cannot articulate precisely how climate change alters our preconceptions and challenge the reproduction of stereotypes, then it fails Eliot's criterion for poetry's social function (1957: 18).

Among Duffy's selection for the *Guardian* there are pieces of both kinds, those that confirm and those that challenge, and to offer a round-up of these here is to give an indicative overview of the kinds of poem that are now being written about climate change. Some depend on simple oppositions between civilization and nature, such as Gillian Clarke's 'Cantre'r Gwaelod', Rachael Boast's 'Silent Sea' and Lachlan McKinnon's 'California Dreaming', or concentrate on bad weather, such as Matthew Hollis's 'Causeway', Michael Longley's 'Storm' and Peter Fallon's 'Late Sentinels', while Jackie Kay's 'Extinction' and

James Franco's 'I was born into a world' are overtly, and in the latter case offputtingly, polemical. Yet there are several poems among the commissions that emerge from or respond to the very difficulty of writing about climate change: David Sergeant's 'A Language of Change' asks confidently – and thus paradoxically with considerable doubt – what it means to write a love poem at the end of the world; while Sean Borodale's 'Scratching for Metaphor in the Somerset Coalfields' explicitly makes an analogy I have proposed in previous chapters – particularly in the work of Bunting and Jones – by thinking about the history of greenhouse gas emissions as a hidden tradition, the fuel like language a medium for storing energies. Although Borodale weaves Coleridge into his poem, he positions a Romantic view of landscape in the sweep of history, unlike Boast, who invokes Coleridge in her epigraph to establish a polar opposition between humanity and the sea.

The historic scope of Borodale's poem is also evident in several other pieces that reach back to the last ice age to imagine how humans might experience a changed climate: Jo Bell's 'Doggerland' overlays two timeframes – the early twentieth century and prehistoric Britain – through the focal image of an antler recovered from the sea; Paula Meehan's 'The Solace of Artemis' envisages the return of the polar bear to Ireland, in a cycle resembling Jones's speculation about future ice ages; and Robert Minhinnick's 'The Rhinoceros' amalgamates Jones's figures of generatrix and craftsman in meditating on the creator of a prehistoric image of the eponymous animal. Taking a different tack, Imtiaz Dharker's 'X' uses the titular letter in various ways to suggest material and sonic interconnections, as Eliot does with water in *The Waste Land* and Jones with wood in *The Anathemata*, and is the only one among the *Guardian*'s selection of offer a perspective that is distinctively not Western.

This whirlwind tour of climate change poetry suggests something of the diversity that the phenomena have prompted in terms of response and engagement, but emphasizes that the lyric mode, as the public expression of the personal, tends to predominate, at least, necessarily, in the most visible examples; attempts are nevertheless being made to address the kind of problems that I have highlighted throughout this book. A thoroughgoing examination of climate change poetry will require the work of a separate study, which could productively correspond with that Adam Trexler has undertaken for the climate change novel in *Anthropocene Fictions*, and could also draw on the work of Sam Solnick's *Poetry and the Anthropocene* (2016). The foregoing discussion is designed instead to frame the context of climate change poetry and attest to the value of more innovative poetics that works consciously in the Modernist tradition, if not to participate in the 'fight' against climate change then at least to help us develop a more sophisticated imagination of the phenomena and not confirm us in our existing positions and politics.

In her 2008 collection *Sea Change*, the US poet Jorie Graham makes just such an adoption of Modernist poetics to engage in the multiple, hybrid phenomena of climate change, and in my concluding analysis of her work, I maintain that there is greater potential in these kind of experimental literary forms for writing our present situation; as I also do that poetry is more successful when it works against assimilation of climate change into traditions of nature poetry and recognizes the value of Modernist artifice and uncertainty of identity.

Sea Change: Modernist poetics and climate change

As well as drawing on the work of Modernist predecessors, Eliot in particular, Graham is in *Sea Change* informed by her own developing practice, which is predisposed towards engagement with the concerns I have discussed in previous chapters. Commenting on Graham's earlier work, Helen Vendler writes that the poet's realization of 'a primacy of the material over the spiritual' is marked by her making 'form mirror the unstoppable avalanche of sensations and the equal avalanche of units of verbal consciousness responding to those sensations' (1995: 106). The result is that 'Formally speaking, "smooth," uninterrupted, unproblematic narration can no longer, for Graham, represent experience, which is forever probing, tentative' (ibid.: 112). That probing quality resonates, for instance, with Jones's 'groping' towards representation at the opening of *The Anathemata* (2010 [1952]: 49).

The particular form sustained throughout *Sea Change* enables various effects as Graham engages with the changing climate. In the title poem, with which the collection opens (Graham 2008: 3–5), this form signifies an uncontainable meteorological potency and agency. The poem begins:

One day: stronger wind than anyone expected. Stronger than
                                    ever before in the recording
                                    of such. Un-
natural says the news. Also the body says it.

(ibid.: 3)

Against the strength of the wind, the force of which is suggested in a blowing-back of sense at the first few line breaks, the poem asserts a feeling of attempted containment through human narrative. The opening 'one day' is the poem's indication of its fictive quality, a 'once upon time' whose abbreviation communicates the urgency prompted by the nature of the story to be told.

Our need for framing discourses is reinforced by more explicit reference to 'the recording' of weather data and 'the news' that interprets it. Notably, these occur ahead of a seemingly tacked-on sensory confirmation, 'Also the body says it'. The priority of media over physical experience signals the ubiquity of discourse in our construction of the world – a marked contrast to Wordsworth's Romantic encounter with the wind early in *The Prelude*. His weather instigates a sympathetic internal response: 'For I, methought, while the sweet breath of heaven / Was blowing on my body, felt within / A correspondent breeze' (1979 [1850]: 31; part I, lines 33–5), which itself trans-forms the classical invocation of the muse into inspirational meteorological phenomenon. Graham's versification in the opening lines of 'Sea Change' confirms that our accounts of weather are only contingent and provisional, however, the forced enjambment of 'Un- / natural' signifying the effort to which we must go to maintain the dualism of unnatural and natural.

While the mediation of phenomenal experience is the ostensible subject of these lines, the shape of Graham's verse makes the poem provocatively rather than evocatively sensory, reminding us of our material presence in the world. As the example quoted above demonstrates, the lines vary in length, so the extent to which our gaze follows them alters accordingly; likewise, our breath if we read them aloud. Graham adopts this distinctive style of versification throughout the book. Poems begin with a line ranged left, sometimes extending across the width of the page but on occasion finishing before halfway; that line is followed in most instances by between one and nine shorter lines that keep a consistent left-hand margin about forty per cent of the way across the page. These are in turn followed by another long line ranged left, then more, shorter lines maintaining the secondary margin at roughly two-fifths of the page width. As in the above quotation, syntax is continuous and the sense of most lines is enjambed. The physical effect of moving from long lines to short is not unlike some of the transitions from prose to verse in *The Anathemata*, and indeed Graham can be read productively according to the methodology Jones outlines in his work's preface: 'I intend what I have written to be said … You can't get the intended meaning unless you hear the sound and you can't get the sound unless you observe the score' (2010 [1952]: 35).[4] Graham's principle resembles Jones's because it emphasizes the distinctive sonic qualities of her form, and represents an engagement with the material rhythms of language that she has been developing throughout her career.

This use of form also enacts a tension between uncontainable material phenomena and the human attempt to impose order on them. Where Motion begins 'The Sorcerer's Mirror' with a pastoral moment of stillness, the momentum of Graham's poem prevents such calm, and she reflects 'how the future / takes shape / too quickly' (2008: 3). The enjambed lines create breaks where we do not syntactically anticipate them, at the same time forcing us to read through them to enact the sense of a future 'taking shape' too quickly for us to control. In several instances, there is a jarring shift in sense or tone; 'how the future / takes shape' by itself can be read as communicating a sense of contented observation, but the subsequent words 'too quickly' snatch that moment from us. The processes of nature cannot be contained by form or syntax. The attempt to do so simply prompts further change: conventional categories are exceeded by enjambed lines when the accelerated passage of time is found to affect 'the state of / being. Which did exist just yesterday, calm and / true. Like the right to / privacy—' (ibid.). As a result, the insistence here that a 'calm and / true' state 'did exist just yesterday' reads as another imposition of human order on a chaotic world, rather than being a simple, nostalgic affirmation of former certainties. Graham's subsequent comparison of this calmness with 'the right to / privacy' thus casts doubt on that right's public assertion of individual selfhood, which is shown to be similarly contingent, an imagined reference point. The poem pursues this tension between human conceptualization – both of the world and of ourselves – and the phenomena that outrun these concepts, because the excess of these phenomena incites a reflexive will to order them on our part.

Human attempts to preserve a humanly defined world in the face of its troubling hybridity remain a preoccupation of Graham's throughout the collection.[5] 'Belief System' (Graham 2008: 45–7), a title that reflects such an anthropocentric vision, opens with an indication of our fictive engagement with the world, and its provisional quality: Graham declares that 'As a species / we dreamed. We used to / dream'. The qualification of tense in the second sentence dispels the definitive quality of what precedes it, but still reaffirms the loss of our capacity to dream. Graham then elaborates on this dream, identifying that it was, in part, characterized by an anthropocentric exceptionalism, because when invoking the mind 'we meant / the human mind. Open and oozing with / inwardness' (ibid.: 45). The sense here seems to be advancing until a line break that drops from 'Open and oozing' to 'inwardness'. This syntactic circularity represents a cultural solipsism, where the reduction of the environment to its category in media and politics forever defers our implication in it: '—we shall put that / off the majesty of the mind / said, in the newspapers' (ibid.).

The poem enacts humanity's attempt to prolong its self-absorption, going on to describe our thought in terms of 'the only / lifetime anyone had' (ibid.). However, the momentum of the lines and the syntactic continuity are hesitant to move on from the isolation this implies, first 'walking' into 'that space', parenthetically stalling for time, and 'then' on into 'the space / of what one meant by one's / offspring's / space. The future' (ibid.). Only after we have tried to formulate time as 'space' and the pause of a full-stop do we move into a definite 'future'. The stop–start rhythm of the lines sets this faltering attempt to manage our transition into that future and slow the arrow of time. But the arrival of the future prompts a further turning to human 'inwardness': 'How could it be performed by the mind became the / question'. In this question, she directly addresses the problem of how to represent phenomena that exist but are beyond the purview of immediate sensation; Graham's characterization of the future as 'this sensation called tomorrow and / tomorrow', an allusion to *Macbeth* (5.5.19), suggests that our engagement with the future can still only be enabled by an engagement with the 'performance' of the past.

Putting the past to use: Recycling Modernist poetry in *Sea Change*

As a poem, and throughout the book to which it gives its name, 'Sea Change' stages a number of references to the canon that help organize and inform Graham's response to climate change. The title's quotation, from *The Tempest*, also alluded to in Graham's poem 'Full Fathom' (Graham 2008: 30–1), signals her engagement with the future through literary tradition. In particular, she recognizes the way that the tradition itself changes through time, just as Eliot does in 'Tradition and the Individual Talent'. In a *Guardian* review of *Sea Change*, M. Wynn Thomas remarks: 'Significantly, this volume's title points us not to the redemptive vision of *The Tempest* but to [it] as ominously refracted through Eliot's *The Waste Land*' (Thomas 2008). Graham's adoption of Eliot sees a further adaptation of this tradition, making work of the Modernist era speak to the present as Pound, Eliot and Bunting all repurposed historic texts for the twentieth century.

One particular concern that *Sea Change* shares with *The Waste Land* is the tension between the human imposition of order and vital, persistent material forces. When 'the future / takes shape / too quickly' in 'Sea Change', it is figured as 'grasses shoot[ing] up, life disturbing life' (Graham 2008: 3); these echo the 'Lilacs' and 'dull roots' from the start of 'The Burial of the Dead' (Eliot 2015 [1922]: vol. I, 55, lines 2, 4). Eliot is able to half-contain natural energies, with the present participles that end the first three lines of his poem – 'breeding', 'mixing' and 'stirring' – creating a cyclical pattern from processes that go beyond the containment of the line, keeping growth temporarily in check. By the twenty-first century, even this momentary equilibrium is impossible, however, and Graham's form signifies the runaway character of material phenomena.

The force of the wind images this quality in both poems as well. In the second part of Eliot's poem, 'A Game of Chess', the wind remains beyond a door, figuring the disturbance of the narrator's interlocutor: '"What is that noise now? What is the wind doing?"' (Eliot 2015 [1922]: vol. I, 59, line 119). As order increasingly disintegrates throughout *The Waste Land*, however, the final section of the poem is exposed to the elements, and we hear 'What the Thunder Said' rather than being able to shut it out. The wind in Graham's poem resembles Eliot's thunder in that it cannot be excluded, exerting its agency from the opening of the poem onwards, and it interjects in the narrator's reflection on the right to privacy to refute the claim that we are unaware of our participation in worldly phenomena: 'consider your affliction says the / wind, do not plead ignorance' (Graham 2008: 3–4). As in Eliot's poem, we strive to enforce the separation that Latour identifies between culture and nature, only to have hybrid phenomena more openly assert their presence.

Counterintuitively, our protestation of ignorance implicates us in change, because the imagined separation of human affairs from meteorological phenomena is what enables such hybrid environments. Graham's 'Sea Change' marks, as *The Waste Land* does, civilization's attempt to create a distinct 'now', a modernity that suppresses its contingent past. This condition is then forced to confront its own artificiality in Graham's poem when she tries to reinforce a physical separation between the human and the phenomenal:

> & farther and farther
> away leaks the
> past, much farther than it used to go, beating against the shutters I
> have now fastened again, the huge mis-
> understanding round me now
> (ibid.: 4).

In *The Waste Land*, despite the attempt to bury it beneath snow and ice, the past repeatedly reasserts its presence; in Graham's poem, the past is imagined not as the dead but as the weather, 'beating against the shutters'. Nevertheless, our resistance to it is still marked by a failed enclosure and separation of human domestic space from nature, though we continue to respond to its 'beating' with repeated attempts to shut it out, 'the shutters I / have now fastened again'. In attempting to accelerate our break with the past, we lose time itself as it speeds out of our control. The 'huge misunderstanding round me now' in Graham is the 'dirty house in a gutted world' of Stevens's 'A Postcard from the Volcano' transposed into our contemporary climatic wasteland, with its own 'spirit storming in blank walls' (1997 [1936]: 129).

Graham's poetry is characterized by such a trespass of the environmental on the territory of the personal. Indeed, Vendler remarks on the materialist direction of her poetics before *Sea Change*: 'The self must now portray itself *in* primary matter; … Yet the indifference of the material universe to our fate makes us hesitate to appropriate its phenomena as adequate symbols of

ourselves' (1995: 125; author's italics). In 'Sea Change', there is evident slippage
of this kind between concept and phenomena when a fleeting thought is
figured as wind off the Atlantic that is in turn personified, 'hissing Consider /
the body of the ocean which rises every instant into / me' (Graham 2008: 4).
There is only a versified – that is, artificial – boundary between the body of
water and 'me', itself neither clearly literal nor metaphorical, in the form of
the line break. With this attention to the personal, Graham enacts Alaimo's
'recognition not just that everything is interconnected but that humans
are the very stuff of the material, emergent world' (2010: 20). So although
existence depends on water, water's significance exceeds this function; like
Eliot, Bunting and Jones, Graham creates a context where water is a signifier
of multiple states, elemental and psychological.

It is in water's very nature to resist reductive characterization; *Sea Change*'s
allusions to *The Tempest* are used to evince a comprehensive breakdown in
the separation between the human and the environmental. In 'Full Fathom',
the narrator seemingly imagines a neighbour into a version of Ariel's song
that begins 'those were houses that are his eyes', before proceeding from these
enclosing houses to their inhabiting 'families', 'privacies' and 'details', on to the
institutional arrangements of 'reparation / agreements, summary / judgments',
through humanity's 'multiplications / on the face of the earth' that lead to our
impact on 'forests' and 'coal seams', and into the 'carbon sinks' that we 'turn
into carbon sources' as they exceed their capacity for storing our emissions
(2008: 31). She brings these phenomena into personal proximity with the
reversal of Shakespeare's formulation: 'Those are pearls that were his eyes'
(*Tem.* 1.2.399) becomes 'those were' / 'that are' in her 'Full Fathom'. The listed
phenomena are then rooted in an experiencing subject, 'his eyes', to become a
vision of a typical individual's implication, through society, with 'carbon sinks'
and 'carbon sources'. But Graham's reversal also serves, rhetorically, to direct
environmental responsibility towards that individual: 'they turn into carbon
sources—his' (Graham 2008: 31).[6]

The ambiguously human figures of *Sea Change*, like those in Eliot,
Stevens, Bunting and Jones, are quasi-objects, difficult to distinguish from
their environments. To stress the physical implication of human beings in
the climate, and the climate in human beings, Graham redeploys *The Waste
Land* again in the poem 'Positive Feedback Loop', using one of Eliot's key

symbols to freight contemporary personal experience with the environmental processes that are beyond both our literal and metaphorical grasp. In this poem, by representing natural processes that occur at the level of Soper's second definition of nature Graham invites us to take Eliot's 'handful of dust' (2015 [1922]: vol. I, 55, line 30) as a tactile model for ocean circulation, and in so doing makes the dust's original spiritual connotations materially manifest.

The awkwardness of the transition from one element to another, from 'dust' to 'water', is suggested by the repeated instruction 'try to', strained in its first occurrence by being broken over the line ending:

In Hell they empty your hands of sand, they tell you to refill them with dust and try
        to hold in mind the North Atlantic Deep Water
        which also contains
contributions from the Labrador Sea and entrainment of other water masses, try to hold a
        complete collapse, in the North Atlantic Drift, in the
        thermohaline circulation
                        (Graham 2008: 42).

The lines run across the page in a manner that demonstrates the difficulty of being able to follow the instruction 'to hold [them] in mind' as we have done in the hand. The conceptually difficult, 'try / to hold in mind', as a result becomes what is physically impossible, 'try to hold a / complete collapse' (ibid.), demonstrating the sensory elusiveness of even a few named climatic phenomena. With the appearance in the poem of named phenomena such as 'thermohaline circulation', or 'convective chimneys' (ibid.), which are neither glossed nor anticipated, Graham is also able to furnish a scientific context by gesturing at the oceanic and atmospheric mechanisms that participate in climate. Not assimilated into the poem, the terms render a necessary poetic and conceptual difficulty.

The complexity of these phrases reflects the effort to which we must go to comprehend the phenomena of our climate; and yet these phenomena remain materially resistant to understanding. The sea of Graham's 'Full Fathom', for instance, is described in the poem's opening lines: '& sea swell, hiss of incomprehensible flat: distance: blue long-fingered ocean and its / nothing else: nothing in the above visible except / water' (Graham 2008: 30). The initial ampersand either invites us to puzzle over the relation between title and first line – between depth and surface – or, if it marks the start of the poem proper, signifies we are always already trying to catch up with a process in progress. By going on to describe the sea's 'hiss' as 'incomprehensible', Graham suggests we presume the noise is something that *should* be amenable to understanding, while by failing to find anything 'visible' other than water, we are forced to identify the absence as 'nothing' (as in the concluding lines of Stevens's 'The Snow Man'). The world's intractability persists in the deferral of syntactic closure in 'Full Fathom'. Colons and ampersands recur throughout the poem, which is also strung through with em-dashes to put off a full-stop until the end of the final line. As a result, the poem accumulates as reformulated statements, stacking up repeated attempts to engage with the world, a kind of groping syntax from which narratives only emerge hesitantly and divergently.

The intellectual grapple to express the sea's ineffability revisits the theme of Bunting's Ode 3 ('I am agog for foam') and more particularly, in its continued recapitulation of engagement, Stevens's 'The Idea of Order at Key West'. In the latter poem, Stevens acknowledges that the sea has an inchoate identity, apart from what the singer sings, and though her song creates a world, it is only the 'world / In which she sang', a solipsism enacted with his line break (Stevens 1997 [1934]: 106). The tension between 'The maker's rage to order' and 'the words of the sea' persists throughout his poem in the very idea that the sea has 'words', and at its conclusion the narrator strains to hear 'Words of the fragrant portals' and 'ghostlier demarcations, keener sounds' (ibid.). In contrast, the sea's noise in 'Full Fathom' is 'incomprehensible' from the beginning, while the narrator of 'Sea Change' can only identify 'syllables untranscribable' in the Latourian 'intermingling' between humanity and phenomena (Graham 2008: 30, 4). Moving beyond Stevens's dialectic of imaginative and phenomenal experience, Graham positions the sea beyond the possibility of even approximate or imposed human understanding. While unable to comprehend the sea, our existence is nevertheless contingent on it, and when Graham alludes to this by describing 'a chorusing in us of elements' in 'Sea Change' (ibid.: 4), her metaphor reverses the personification of the world through song in 'The Idea of Order at Key West'. Graham is closer to Bunting in Ode 3 in making human identity dependent on the sea.

The musical terminology of 'chorusing in us' recurs in subsequent poems in Graham's collection. The lines 'Who is one when one calls oneself / one? An orchestra dies down' (Graham 2008: 42) and 'dead gods … turn the page for / us. The score does not acknowledge / the turner of / pages' (ibid.: 45), voiced as asides in their respective poems,[7] also express our interdependency, figuring the self as one of many in a concerted musical effort. Like the orchestra, humanity can create an harmonious world, albeit only through the provisional, imaginative medium of metaphor. This creation compares with the way the trumpet heralds and makes comprehensible the weather in poem VIII of Stevens's 'Credences of Summer' (1997 [1947]: 325), and indeed meteorology *is* music in Graham's 'The Violinist at the Window, 1918': from the viewpoint of the titular figure, imagined from Matisse's painting, 'the mind is hatched and scored by clouds / and weather' (2008: 33).

The extension of musical motifs across Graham's poems also enacts that context of mutual and multiple creation. It is a cumulative effort, however – Stevens's 'personage in a multitude' (1997 [1947]: 326) perhaps, rather than his solitary singer at Key West – from which this metaphorical harmony emerges. When we revert to the conception of ourselves as individuals, the orchestral effect goes unrecognized: the individualism of instead 'calling oneself one' in 'Positive Feedback Loop' for instance means that the music 'dies down' (Graham 2008: 42). Having shown individualism to be implicated in environmental change, in particular in 'Full Fathom', Graham intimates that we aggravate that change when we behave as individuals rather than engaging with it through a collective, orchestral understanding of human behaviour of the kind that the Anthropocene must entail.

In 'Credences of Summer' and 'The Idea of Order at Key West', Stevens proposes that music serves as an attempt to engage with the world, but because the world always exceeds our songs of it, we are forever required to recapitulate these engagements. Stevens presents the imagination as the dynamic operator in this equation, but Graham has to recapitulate her imaginative projections under the water in 'Sea Change' paratactically to pursue dynamic, accelerating environmental processes:

> at the very bottom of
> the food
> chain, sprung

> from undercurrents, warming by 1 degree, the in-
>
>   dispensable
> plankton is forced north now, & yet farther north
>
>   (ibid.: 4).

The line breaks signify human imposition on the flow of environmental process, the only way in which we can render the 'indispensable' dispensable, or interrupt (with a page's end) the subsequent 'un- / interruptible slowing of the / gulf / stream' (ibid.: 4–5). The emphatic momentum of these lines signals the processes that drive through them, so when Graham invokes 'undercurrents, warming by 1 degree,' we cannot linger on them but proceed directly to their effect. Climate change is thus something that demands our consideration not just because its terminology is invoked in these poems, but because the poetry enacts the climate's participation in all of our other concerns.

As the poem draws to a close, it marks a movement from environmental to personal. Graham writes 'so that I, speaking in this wind today, out loud in it, to no one, am suddenly / aware / of having written my poems, I feel it in / my useless / hands.'[8] By 'speaking in this wind', she is giving it voice, speaking 'in' its form and responding to our need to sing the world, as at Key West; but she could equally well be speaking in*to* it, and 'out loud' at that, to overcome the material quality of its noise. In both cases, it suggests another agency at work that is irreducible to human intention or representation, and Graham's reference to 'useless hands' implies that this agency contributes to the creation of the poem through her; the narrator is 'suddenly aware of having written [her] poems'.[9] Rather than seeking to describe her surroundings,[10] she gives them a voice: 'quicken / me further says this new wind' (2008: 5). The verb dramatizes the process of 'quickening' as bringing to life in language, but also implies human responsibility for accelerating and intensifying weather patterns, and their 'uninterruptible' course. Graham shares Kerridge's recognition that 'Lyric poetry that uses an "I" persona also has difficulty with these perspectives, having to bring them within the frame of a dramatized personal consciousness' (2013: 352). As a result, while she includes such a persona in the poem, it is neither the entirety nor even the centre of the piece. 'Sea Change' makes a Stevensian acknowledgement of the limitations of lyric selfhood, because that self is characterized by its repeated failure to manage the world.

The narrator of 'Sea Change' makes a final attempt at order, but the world proves beyond her control. As Motion does in 'The Sorcerer's Mirror', Graham's narrator returns to her garden at the end of the poem, having situated it in a problematic global climate: the personified wind observes 'your / best young / tree, which you have come outside to stake again' (Graham 2008: 5). Yet, while Motion turns away from the environment as his narrator returns to the house at the end of his poem, the longer scope of her collection gives Graham the opportunity to show what such turning away might mean in the changing climate. As 'Positive Feedback Loop' moves towards its end, she envisages 'us in The Great Dying again, the time in which life on earth is all but wiped out / again—we must be patient—we must wait—it is a / lovely evening, a bit of food a bit of drink' (ibid.: 44). The bathetic movement from extinction to dining arrangements communicates both the simultaneity of everyday living and ecological collapse and the ineffectualism of the sensual lyric self in that context, what Morton describes as 'The gratifying illusion of immersion in a lifeworld [that] provides yet another way to hold out against the truth of global warming' (2010b: 10).

'No Long Way Round'

The final poem of *Sea Change*, 'No Long Way Round', marks some subtle departures from the form that characterizes the collection's other poems. The syntax still circles and qualifies in response to natural phenomena – 'Evening. Not quite. High winds again' (2008: 54) – but the full-stops make it terser than earlier pieces. In this context, Graham explicitly confronts the paucity of prior modes of understanding in the face of climate change, recognizing their obsolescence in this context: she invokes 'telling / the truth' as a duty of the past, only to say that 'We / liked / the feeling / of it—truth—whatever we meant by it—I can still / feel it in my gaze, tonight, long after it is gone, that finding of all the fine discriminations' (ibid.). This 'finding of … discriminations' echoes Stevens's lines in 'Variations on a Summer Day' that 'The difference between air / and sea exists by grace alone' (1997 [1942]: 215) – and Graham goes on in her poem to demonstrate what it means to live without this capacity in a more striking formal divergence from the pattern of the rest of her book.

In two passages of 'No Long Way Round', the verse clumps into a pair of paragraphs resembling prose. The first of these reads:

> It is an emergency actually, this waking and doing and
> cleaning-up afterwards, & then sleep again, & then up you go, the whole 15,000
> years of the inter-
>> glacial period, & the orders and the getting done &
> the getting back in time and the turning it back on
>>> (Graham 2008: 55).

The day-to-day routines are already described as an 'emergency', but their rhythm carries us through the verse paragraph, 'waking and doing'. The transition represents the way in which the discourse of 'emergency' becomes normalized in the everyday, whether in an attempt to ignore it or simply diminish its force. Graham builds up momentum in the first verse-paragraph before suspending it in the elongated line break. Should we mark that gap by 'reading' silence as a pause, it reminds us of the brevity of the current interglacial;[11] if we mark it by holding the reading breath, though, we realize the physical difficulty of even one unspoken line. By either scale, the effect reminds us of our physical implication in the world. Graham's resumption with a further prose-like stanza creates an illusion that things are close to normal, but the poetic interruption serves to express the contingency of our quotidian lives.

As with preceding poems in the collection, 'No Long Way Round' draws in with a moment of lyrical meditation: 'You have your imagination, says the evening. It is all you have / left, but its neck is open, the throat is / cut, you have not forgotten how to sing, or to want / to sing' (ibid.). Notwithstanding that the cut throat is more violent than any of Stevens's images, it doubly affirms, as his poems do, a failure to articulate an imagination that is adequate to the world, along with our continual drive to 'to sing' our songs of the world anyway. Yet Graham shows how the story we would tell is confined to its profoundly human significance, because it comprises

> how you met, the coat one wore, the shadow of which war, and how it lifted,
>> and how peace began again
>> for that part of
> the planet, & the first Spring after your war, & how "life" began again, what
>> normal was—thousands of times
>> you want to say this—normal—
>>> (ibid.: 56).

The story begins personally, and even when it attempts to broach a more global scale it tries to contain it, at first to the body with 'the coat', then limiting it to 'that part of the planet' and 'your war', the self-consciousness of what constitutes 'life'[12] – and finally with the desperation to restate normality. If our fictions remain as local as this, however, then they are likely to end, with us, sooner rather than later.

Graham's achievement in *Sea Change* is to use the poetic medium itself to mark our inability to contain the world's significance in terms we understand. Thinking of time as space, as she invites us to do in 'Belief System', we will find no long way around climate change, no saving diversion, given that it threatens to telescope transformations of a geological magnitude into generational time – exacerbating existing phenomena beyond the planet's ability to accommodate them, let alone ours. In the collapsing of both Shakespeare and Eliot into her sea change, she dramatizes the exponential acceleration of our traditions, both those we consciously maintain and those we unconsciously perpetuate, in the creation of a fleeting present moment.

Conclusion: The New Poetics of Climate Change

Rereading Modernist poetry in the light of climate change can enable us to respond to the limitations of ecocriticism that I outlined in my first chapter, and its neglect, until very recently, of early twentieth-century literary aesthetics. The Modernist poets whose work I have considered engage more incisively with the Romantic tradition, through their creative interrogation of it, than do critics seeking to recover a hedged-in Romanticism as a paradigm for our relation with nature. Wallace Stevens and Basil Bunting in particular are concerned with the way that our relation with nature has altered over time, and seek its imaginative root rather than seeing it as a way to restore us to an idyllic wild world. Our understandings of nature are drastically modified by the complexities and contingencies of urban living as these intensify in the nineteenth and twentieth centuries, and such processes are addressed in the poems of both Bunting and T. S. Eliot. They bear witness to the strain that the city – as a manifestation of civilization and capitalism – places on our relations with nonhuman phenomena, and the way that culture cannot resist the materially deleterious effects of exacerbated environmental change. Yet David Jones's imaginative scope in *The Anathemata* shows too that we needn't reject Romanticism out of hand when considering twentieth-century poetics; rather, that we need to situate it in a context problematized by subsequent scientific findings. Because Modernist writing occurs at a historical moment between Romanticism and the manifestation of anthropogenic climate change, the traces of literary past and environmental future are entangled and exposed in its work.

Another benefit in rereading Modernism from an ecocritical perspective is that, by tackling texts not topically related to environmental crisis, we begin to understand how far our patterns of thought have to alter if we are to confront the full implications of climate change. Reading climate change into the preserve of canonical Modernism shows how extensively the phenomena can destabilize our patterns of thinking: it cannot be reduced to its iterations in political or environmentalist discourses, neither, entirely, to its scientific models or analyses. Not only the causes but the effects of climate change are entangled in human practice, and by imagining those consequences in climatic 'fictions', along the lines that Stevens entertains, we can develop the imaginative resources that will inform our cultural adaptation.

Modernist poetics has particular further value in the consideration of climate change because it enables us to read the increasing complexity of unsituated environmental risk identified in the sociology of Bruno Latour and Ulrich Beck. Modernism's engagement with an emerging globalism is evident in Eliot's and Bunting's metropolises, in Stevens's multiply situated visions of 'The Planet on the Table' and in the temporal and geographical scope of *The Anathemata*. The more innovative and open forms employed in much of this work can be read as expressing an understanding of human implication in forces beyond our ordering or control, forces which are yet sensitive to our interference. Even the formally cautious Stevens is elliptically restive in his poems and refuses to endorse a stable sense of self or of the world. The Modernist use of motif accentuates the resonance of the objective particular with the abstract general, attuning our imaginations to the environmental significance of our individual experiences. As such, they offer an expression of the theoretical tenets of material ecocriticism: the poetry becomes various sites of interaction between intentional and unintentional agencies – cultural and phenomenal, conscious and unconscious, authorial and futural. By not being topicalizations of 'environmental crisis' or 'climate change', the work can explore the forces and principles that contribute to its emergence across the twentieth century, rather than its symptoms. The work's reception in a changed climate also marks literature's material persistence, its resistance to determination by the criteria of historical context, as an aesthetic modelling of unintentional phenomena.

Rather than offer here a historicist reading of the Modernists' own under-standings of and engagements with ecology, of the kind found in Jeffrey Mathes McCarthy's *Green Modernism* and Joshua Schuster's *The Ecology of Modernism*, what instead I have been developing is a kind of Modernist ecocriticism, reading the discipline through the poetics that began to emerge 100 years ago. My reinterrogation of key Modernist texts in a contemporary context has thus brought out different qualities of the hybrid phenomena of climate change; the plethora of responses reflects our multiple vectors of entanglement with the phenomena, and the range of agendas or fictions to which we assimilate it.

The readings I have offered in this book of work by Eliot, Stevens, Bunting and Jones together represent three key ecocritical qualities in Modernism: first, that Modernist work offers a fuller and more engaged reading of Romantic relations with nature as they are altered in the industrial era than does early ecocriticism; second, that a consideration of this work moves ecocriticism beyond its reliance on texts that are largely or wholly concerned with nature or environment as topics, providing it with greater nuance and unexpected insights; and third, that it offers us a way of reading our environmental entanglements as they become increasingly complex with anthropogenic climate change. There is one further quality that a reading of Modernism then offers, which I have explored in Chapter 6: specifically, how its aesthetics represents an alternative, unconsidered tradition for the writing of environmental entanglement, a Modernist poetics of climate change. This provides a valuable way of tracing the complexities of the phenomena, and resists its reduction to a collection of tropes. Further work remains to be done in this regard, and most helpful would be a comprehensive survey of climate change poetry to ascertain the full range of modes deployed; this could be modelled on the survey of climate change in Anglophone novels conducted by Adeline Johns-Putra and Adam Trexler, which informed the latter's *Anthropocene Fictions*.

At the last, we should bear in mind that climate change is not simply the transition from one fixed state to another but a continuous process that operates according to the principles of Soper's second definition of nature, exacerbated in the past few centuries in particular by human activity. The readings I have offered will, and should, be themselves subject to change. Indeed, accommodating this process of change should be constitutive of ecocriticism, on Richard Kerridge's analysis:

> Ecocritical responsibility consists in accepting that the existence of a large expert majority for a view constitutes a form of probability that the view is correct—the only form of probability a non-scientist ecocritic can scrupulously acknowledge. If the majority view changes, then the ecocritic has a responsibility to change accordingly, without needing to feel guilty of previous

misjudgement, since to do so would imply a capacity to make expert judgements upon the data. (2013: 349)

Kerridge's remarks can help us distinguish the role and responsibility of a climate change criticism from a climate change poetics. My readings here are necessarily contingent on the network of critical and scientific understandings contemporary to my writing this book, and these will change with time, as Kerridge observes. As a work of criticism, this book is an act of explication and interpretation, communicating a particular understanding. It is therefore designed to have clarity and to eschew connotation for denotation, so it is contingent on the climate of criticism and research that produced it.

The poetics of climate change, however, consists in both the poetry and climate change, the hybrid that these two terms represent. In these are entangled the forces and phenomena so criticism's task remains, in Hulme's words, 'to reveal the creative, psychological, ethical and spiritual work that climate change is doing for us' (2009: 326). Climate change poems must be sufficiently poetic to change with the climate, outrunning critical pronouncements on them as the phenomena themselves do. The poems I have studied in this book, along with those that are being and have yet to be written – perhaps, by extension, any poems – will remain articulations or sites of interaction between the forces described, exposing networks of agency and making legible the entangled processes of climate change.

Notes

Chapter 1: Climate Changes Everything

1   I will use the capital-N form, 'Nature', in this book when I invoke an uninterrogated nostalgic, green vision of the term.

2   Adam Trexler identifies this problematic location of climate change at the poles, writing that, 'Intuitively, it would seem that a global process like climate change could be described anywhere'; yet 'In the first decade of the twenty-first century, media coverage of climate change was dominated by images of the Arctic and Antarctic' (2015: 78, 80).

3   As I will go on to show, climate change resists attempts to understand it as a single phenomenon, so I will refer to it as a (network of) phenomena throughout this book.

4   His most marked divergence from the sonnet form, the addition of a fifteenth line to the second poem, might correspond to the melting of ice 'in deeper currents and quick, chaotic flow' (Motion 2009b), or it might be a requirement of the sequence's musical setting. But this slight variation cannot be said to present a significant formal or modal innovation, and indeed depends on the form remaining otherwise stable to achieve what minimal effect it has.

5   Robert Markley refers to this phenomenon as 'anthropogenic history' – the sense of time that human existence entails – and its implied opposite is a 'time that transcends and beggars human experience' and yet which 'can be conceived only differentially, paradoxically, in its relation to phenomenological perceptions of time and experience' (2012: 46, 52).

6   Bate's use of the term 'global warming' reflects its more general use in the late 1980s and early 1990s. Lorraine Whitmarsh explains that:

> Since the 1980s, the term 'global warming' has been commonly used to describe the impact on climate of increased levels of greenhouse gases linked to human activities. While the 'warming' metaphor may have been effective in capturing the public's imagination about this global risk, it obscures the complex and potentially devastating range of effects resulting from what is more commonly referred to amongst scientists as … 'climate change'. (2009: 403)

To distinguish between my use of the terms, I will largely employ 'climate change' to signify the network of phenomena that are of concern in this book. Those phenomena are both human – cultural, economic, industrial, agricultural – and biophysical – atmospheric, oceanic, solar, botanical and so on. I will also endeavour, for consistency's sake, to use the term 'climate change', and in contrast only refer to 'global warming' where meaning demands the distinction between increasing average terrestrial temperatures and other effects of the changing climate – polar ice loss, species migration or extinction, seasonal shift and so on. However, given the necessity of regular references to 'climate change', I may use 'global warming' as a synonym in places to ease the burden of repetition on the reader.

7   Bate omits reference to this in the reworked version of the essay that appears as 'Major Weather', Chapter 4 *The Song of the Earth*; I have drawn on the essay rather than the book here because of this assertion about 'Global Warming Criticism'.

8   As a result of these conditions, Tim Flannery points out that 'agriculture commenced ... around 10,500 years ago in the Fertile Crescent' in what is the present-day Middle East (2006: 53) – that is, civilization only emerges when the conditions enabled it do so, and, reflexively, human beings recognize these qualities as 'natural'.

9   I have proposed as much in the article 'Climate Change and the Individual Talent' (2014).

Chapter 2: A New Climate for Modernism

1   Bunting's biographer Richard Burton records that the younger poet, for instance, 'berate[d] Pound for his failure to acknowledge fascist atrocities such as Guernica and for ... his anti-Semitism' (2013: 261).

2   The term derives from Buell, who argues for an 'environmental unconscious' that is simultaneously 'the limiting condition of predictable, chronic perceptual underactivation in bringing to awareness, and then to articulation, of all that is to be noticed and expressed' and 'a residual capacity ... to awake to fuller comprehension of physical environment and one's interdependence with it' (2001: 22). But it will not suffice simply to elevate our awareness of the immediate environment and take notice of the changes wrought by the changing climate because this is to treat it simply as something environing or surrounding us rather than a hybrid of cultural

and natural phenomena. Timothy Morton's alignment of environment and unconscious problematizes literature's very ability to do so on Buell's model: 'Nobody likes it when you mention the unconscious, ... because when you mention it, it becomes *conscious*. In the same way, when you mention the environment, you bring it into the foreground. In other words, it stops being the environment' (2007: 1; author's italics). A literary text need not foreground the environmental to attend to its agency, though there are still ways that the environmental unconscious can be registered poetically, as I will suggest in Chapter 4.

3   I give fuller consideration to the implications of climate change for the aesthetics Eliot sets out in that essay in 'Climate Change and the Individual Talent: Eliotic Ecopoetics' (2014).

4   I acknowledge Kerridge (2007: 131) for reminding me of the pertinence of Lanchester's remark.

5   Leonard M. Scigaj, for instance, claims that 'changes in personal behavior and voting patterns' to more responsible positions 'are more likely to appear after the reader has altered his or her perceptions about how humans ought to relate to the environment', and it is the function of what he calls 'sustainable poetry' to occasion such changes in perception (1999: xiv).

6   Alaimo herself examines Muriel Rukeyser's *The Book of the Dead* (1938), which the critic describes as 'a modernist poem of material documentation, folding scientific, medical, and legal evidence into a literary form' (2010: 44–5).

7   Incidentally, the formulation of the question on this occasion, 'how should I begin / To spit out all the butt-ends of my days and ways?' (ibid.: 7, lines 59–60), makes smoking typical of the persona's habits, itself a mutual pollution of body and local environment and a trans-corporeal transgression of the supposed frontiers of the self.

8   Evolutionary and climatic change would at least coincide at that point, because the effects of human activity across what is, geologically speaking, a very short span of time will persist for many millennia more, and in all probability be responsible for a sixth planetary extinction event. Frederick Buell writes: 'what human society, at present, is extinguishing at so unprecedentedly dizzying a rate can be repaired only in time frames at least several times longer than the evolutionary span of the human species' (2004: 107). This has prompted some to advocate for the instantiation of a new geological epoch, the Anthropocene, in which humanity represents the dominant geological force. I will explore this concept in greater detail in Chapter 5.

9   Although such resonances are characteristic of literary texts, this quality

is amplified by what I have earlier identified is Modernism's self-reflexive textuality and fictionality (see p. 32 above).

10 Environmental change is already part of the poem's particular history, as Adam Trexler (2011) points out: he discusses how *The Waste Land* was composed when efforts to intensify British agriculture in the years after the First World War were having an impact on the British landscape.

11 '[T]he idea of a normative climate', Martin McQuillan writes, is 'derived from the idea that a change in climatic conditions would constitute a crisis for the human race' (2012: 275): 'to identify an event as a crisis is always to ontologize it and submit it to the model of the crisis that would explain it and domesticate it', and 'The naming of a crisis in the present works to mask that history and to neutralize it, giving it form and therefore a program and calculability' (ibid.: 274). We may also note the intentionality – and fussy particularity – of the crisis that prompts the Prufrockian hesitation: 'Should I, after tea and cakes and ices, / Have the strength to force the moment to its crisis?' (Eliot 2015 [1917]: vol. I, 7, lines 79–80).

12 Eliot's reference can be found at *AC* 2.2.202.

13 Jorie Graham does much the same in her own adoption and adaptation of the motifs of *The Waste Land* in *Sea Change*, as I will demonstrate in Chapter 6.

14 As does Pound: North points out that in one context, his translation of Cheng Tang's motto offers 'make it new as the young grass shoot' (North 2013).

15 Trexler's essay on the economic context of the poem describes the environmental change resulting from the 'unprecedented control' that the British government exercised over agriculture during the First World War, 'setting production targets for farmers, fixing prices and expanding by 30 per cent the land under cultivation' (2011: 278). This resulted in 'the distinctly modern crisis of overproduction' after the war, and Trexler notes in contrast that the 'Victorian public moralists that Eliot lectured on in his [adult education] extension courses were inspired by medieval agriculture's interdependence with natural abundance, cyclical and social order. Post-war overproduction turned these idylls into nightmares' (ibid.: 279).

16 Nigel Clark notes that 'The abrupt climate change thesis speaks of thresholds which, once passed, leave climate systems tipping rapidly and irretrievably into alternative states' (2010: 32).

Chapter 3: Wallace Stevens's Fictions of Our Climate

1   Two of the four uses of the word 'climate' by Stevens, as identified in the
     Online Concordance to Wallace Stevens' Poetry (www.wallacestevens.com/
     concordance/WSdb.cgi [accessed 22 October 2016]), occur in successive lines
     of the sixth poem of 'The Sail of Ulysses' (1997 [1954]: 465): first in a simile in
     which it signifies a generative environment that creates particulars, and second
     as a metaphor for mental states.

2   Dates given for poems discussed in this chapter will be those of their first
     appearance in a collection, unless otherwise stated.

3   In her reading of Stevens's 'The Man Whose Pharynx Was Bad' (Stevens 1997
     [1931]: 81), Maggie Kainulainen considers 'an experience of extreme cold' as
     something that would 'disrupt his [i.e. the title figure's] rote routine, and make
     decisive action possible. The poet imagines with utter clarity what this intrusion
     of the natural world into the inner world of consciousness might feel like … yet
     the poem ends despondently' (2014: 109). That is to say, there is a yearning for
     cold inverse to the diminution of self in 'The Snow Man'.

4   Human implication in the phenomenal world, culturally and physically, is
     reaffirmed in the subsequent, third poem of 'Credences of Summer'. The figure
     of an old man, nearing death, reaches the boundary of human comprehension,
     but the sense remains that phenomena will always exceed even a lifetime's
     understanding. Any identification of selfhood in the external world is then, as it
     is here, the pathetic fallacy acknowledged as such, the imagination projecting its
     own mortality on to its natural surroundings.

5   Basil Bunting's reference to Judgement Day in *Briggflatts* is explicitly religious,
     deriving from Islamic lore as Don Share explains (Bunting 2016: 348–9 n.
     to part III, lines 90–5); the poet depicts the angel 'Israfel, / trumpet in
     hand / intent on the east'. Bunting does retain the association between the
     announcement and weather, however, with the angel's 'cheeks swollen to blow'
     and 'whose sigh is cirrus' (Bunting 2016 [1965]: 53; part III, lines 90–3).

6   The only other occasion on which Stevens uses the word 'climate' apart from
     here and in 'The Sail of Ulysses' is in 'On the Way to the Bus' ([n.d.] 1997: 472),
     where it is used partly as a recapitulation of and partly as a complement to the
     time of year, an ever-broader frame through which to see a particular part of the
     day, again suggesting something after which the imagination is always reaching.

7   Judith McDaniel argues that 'in all of his work, Stevens expresses a distinctively
     "scientific" imagination' (1974: 223). By her understanding, 'Science in the
     twentieth century, particularly physical science … has gone far beyond the

"commonsense" rational approach of the last three centuries to a highly speculative, imaginative approach' (ibid.: 222). The theoretical ambition she identifies as common to Stevens and his scientific contemporaries is problematized by Szerszynski's account of instrumental climate modelling; the scientific imagination is directed towards results, whereas the poetic imagination reflexively examines itself.

8   The statistician George Box declared that 'ALL MODELS ARE WRONG BUT SOME ARE USEFUL' (1979: 202; author's capitals). Successful statistical modelling of any phenomena requires something Box calls 'parsimony', that is, a small number of parameters (ibid.), and while the countless inputs and impacts of climate change make the process of climate modelling necessarily tricky, poetry as a form tends towards a parsimony of expression, making the fewest words express the most, and the most precisely.

9   A more accomplished examination of this co-creation is to be found in Stevens's later 'The Idea of Order at Key West' (1936), a poem discussed in my next chapter in comparison with an ode by Basil Bunting (see pp. 116–18).

10 A similar elision is apparent in poem XV of Stevens's 'Variations on a Summer Day' (1997 [1942]: 212–15), where 'The last island and its inhabitant' remain 'Until the difference between air / And sea exists by grace alone'. As the distinction becomes increasingly slim, it demonstrates the discrepancy between human practices of naming and phenomenal process such as the hydrological cycle. With the material emergence of climate change, distinctions such as these are more and more tenuous. Stevens positions this recognition with 'The last island and its inhabitant', as though the near-continuity of air and sea has submerged all other islands. Read in our contemporary context, the poem provides a salutary reminder of what 'grace' there might be that prevents final inundation.

11 We might regard this as a definition of sustainability *avant la lettre*, given its concern with the survival of human institutions in the face of unprecedented change.

12 Clark here draws on a point made by David Wood when he discusses everything that we need to have represented by what stands before us:

> Suppose I look out the window—what do I see? A tree. There it is. It is there in front of me, as visible as I could want … [But] the life of the tree, the living tree, the tree of which we glimpse only a limb here, a trunk there, or views from various angles, this temporally extended persisting, growing tree, is invisible. (2005: 152)

This contrasts with a(n in)famous observation in early ecocriticism by Lawrence Buell; asking us to imagine him looking up from the writing of the book at 'the

grove of second-growth white pines that sway at this moment of writing …
forty feet from my computer screen' he seeks to affirm the existence of a world
beyond the text, in contrast to 'The forest of American scholarship', which 'is
a forest where treeness matters but the identities and material properties of
the trees are inconsequential' (Buell 1995: 10). His gesture beyond the page
becomes totemic in ecocriticism for the idea of nature as a simple object of
reference, even though his trees' materiality fixes such nature in particular
entities. Wood actually does more to offer *some sense of the environment as a
process rather than a constant or a given* that Buell requires of environmental
writing (ibid.: 8; author's italics), by considering the tree not as a fixed object
but an organism growing through time. It is Wood's contention that such
consideration can 'activate and reactivate the complex articulations and
relations of things, restoring through description, through dramatization, a
participatory engagement (bodily, imaginative, etc.) with things' (2005: 153).

13   Waste disposal extends beyond the intentional act of binning and landfilling
litter. As Miller observes, it also includes 'all the carbon dioxide in the
atmosphere from automobiles, coal-fired electricity plants, and other sources
that is a chief contributor to global warming, … all that methane from domestic
cows and from landfills, … all the smoke from forest-clearing' (2010: 84).

14   Holly Stevens recalls 'a vast stretch of barren land that people used as a dump …
On this lot a man, seemingly coming from nowhere, built his home. A glorious
shack, made of all the appropriate junk that could be found, with even a
chimney: only when we noticed smoke coming out did we realize someone was
living there' (1971: 652). The Stevenses detect the presence of another human
from the emissions he creates rather than his own person, demonstrating that
scale and perspective are a crucial determinant of whether we focus on human
beings or the environment: the poet's position at his window enables a more
Ariel-like view of the state of the atmosphere.

15   Morton makes a useful assertion that connects such persistent waste with
climate change: 'capitalism creates things that are more solid than things ever
were. Alongside global warming, "hyperobjects" will be our lasting legacy.
Materials from humble Styrofoam to terrifying plutonium will far outlast
current social and biological forms' (2010a: 130).

16   This distinction will also be evident in the poetics of entropy I identify in the
work of Basil Bunting in Chapter 4.

17   As the dump of 'The Man on the Dump' is an image of an artificial site, so too
is the greenhouse. Also like the dump, the greenhouse may be inspired by a site
in Stevens's neighbourhood – Holly Stevens mentions 'The greenhouses [that]
were on our route' to Elizabeth Park in Hartford (1971: 656). In both of his

poems, Stevens abstracts particulars from his own environment to make them into resonant images.

18　This is the translation provided by editors Frank Kermode and Joan Richardson in their note to the poem (Stevens 1997: 1002).

Chapter 4: Basil Bunting and Nature's Discord

1　Note how the transition Hatlen describes from abstract to particular is also the scope of Stevens's 'The Planet on the Table'.

2　This is akin to another of Pound's dicta: 'As regarding rhythm: to compose in the sequence of the musical phrase, not in sequence of a metronome' (2009 [1918]).

3　Bunting cryptically says in his note on the poem: 'Gibbon mentions its effects in a footnote' (2016: 537). Gibbon's footnote itself reads: 'Lycopolis ... has a very convenient fountain, "cujus potû signa virginitatis eripiuntur"' (1994 [1781–8]: II, 64 n.112.) The Latin translates, 'on drinking which, the signs of virginity are torn'.

4　This is most notable in section III of *The Well of Lycopolis*: 'Can a moment of madness make up for / an age of consent?' (Bunting 2016 [1935]: 23; part III, lines 37–8) spoofs Eliot's 'The awful daring of a moment's surrender / Which an age of prudence can never retract' (Eliot 2015 [1922]: vol. I, 70; lines 403–4). Bunting also vulgarizes 'Twit twit twit' (Eliot 2015 [1922]: vol. I, 63; line 203) in 'tweet, tweet, twaddle, / tweet, tweet, twat' (Bunting 2016 [1935]: 23; part III, lines 19–20).

5　Such a regional emphasis can be seen, for instance, in the first of Bunting's own notes on the poem 'Northumberland differs from the Saxon south of England [and] Southrons would maul the music of many lines in *Briggflatts*' (Bunting 2016 [1965]: 538). While Bate tends to endorse Bunting's regionalism, referring to *Briggflatts* as a 'Northumbrian poem ... in which identity is forged in place' (2000: 234), Tarlo rightly identifies that Bunting 'chooses (some would say disingenuously) to emphasize the rural and local aspects of a poem which in fact ranges the world and time' (2000: 152).

6　This is consistent with an understanding of systemic ecology that emerges in middle of the twentieth century. Donald Worster outlines the conceptual value of the ecosystem in *Nature's Economy*: 'Using the ecosystem, all relations among organisms can be described in terms of the purely material exchange of energy and of such chemical substances as water, phosphorous, nitrogen, and other nutrients … These are the real bonds that hold the natural world together; they create a single unit made up of many smaller units—big and little ecosystems' (1985: 302). The 'ecosystem brought all nature—rocks and gases as well as biota—into a common ordering of material resources. It was more inclusive, paradoxically, because it was first more reductive' (ibid.), not unlike the principles on which climate modelling is based (see Chapter 3, pp. 70–2).

7   A recent study demonstrates the dangerous impact that such an understanding can have on the environment. According to a paper by Kim Naudts et al. 'Europe's forest management did not mitigate climate warming' in the last three centuries, because the conifers with which humans have replaced broadleaved tree species in that time have tended to increase heat retention, offsetting their uptake of carbon. Naudts et al. find that 'since 1750, in spite of considerable afforestation, wood extraction has led to Europe's forests accumulating a carbon debt of 3.1 petagrams of carbon' (2016). By regarding trees as an interchangeable resource instead of particular kinds of tree as integral to continental ecosystems, we have been unable to account for their agency in our environment. The study should also prompt caution in that it reminds us we cannot focus alone on reducing levels of atmospheric carbon as a straightforward 'fix' for climate change, because measures intended to do so will have their own unintended impacts.

8   Note, too, that Dempster also suggests forests are characterized as autopoietically mechanistic so they can be economically co-opted by the 'common "harvesting" mentality' (2007: 105).

9   Nigel Clark remarks on 'the element of excess that attends the unearthing of a previously inaccessible fire source, or what is effectively the making present of past solar energy' (2012: 270).

10   In the context of part II of *Briggflatts* – in particular the line 'There is a lot of Italy in churchyards' (Bunting 2016 [1965]: 47; part II, line 106) – there is an apparent allusion to Keats's Italian gravestone, inscribed 'Here lies one whose name was writ in water' (Motion 1997: 564). One possible influence of Keats's choice of epitaph is Shakespeare and John Fletcher's *King Henry VIII*, 'Men's evil manners live in brass, their virtues / We write in water' (4.2.45–6). Motion says Keats 'had devised an inscription which adapted the translation of a Greek proverb', and interprets it as meaning his poetry 'was [now] part of nature – part of the current of history' (1997: 565). Oonagh Lahr, whose scholarship Motion cites (ibid.: 604 n.3), includes Shakespeare and Fletcher's lines among the possible inspirations for Keats's epitaph. However, her main argument is that although 'The bitter epitaph Keats devised

for his own grave is sometimes supposed to derive from a line in Beaumont and Fletcher's *Philaster*, the English sources are indebted to 'a proverbial expression in ancient Greek' (Lahr 1972: 17). A. J. Woodman rebuffs Lahr's assertion, claiming that 'Since "writing in water" occurs numerous times in English poets—in Shakespeare, among others—it is *a priori* more likely that Keats's mind was not on classical literature at all' (1975: 13). In either event, Bunting's lines resonate in a long tradition, situating anonymous description in the current of natural history. These resonances extend the significance of Bunting's 'name cut in ice' so that it includes reputation, inscription and language; what is at stake in the transition between ice and sea, then, is the trace of humanity itself.

11   In respect of the autobiographical basis of Bunting's poetry, Richard Burton elaborates on his relationships with Peggy Mullett and Mina Loy, the dedicatees of Odes 3 and 17 respectively (2013: 131–2, 189).

12   This distinguishes the selfhood of Bunting's poems from the authoritative, autopoietic entity that the self may appear to be in Pound, or aspire to be in Eliot.

13   While there is an historical resonance here, given Bunting's service in the RAF and diplomatic corps in Persia during and after the Second World War detailed in the third chapter of Burton's biography (2013: 269–336), this part of the poem lacks the closer resemblance of the poet–hero in parts I and II to the biographical Bunting.

14   In *Briggflatts*, Bunting incidentally fulfils other of Legler's criteria for ecofeminist writing, specifically: '1. "Re-mything" nature as a speaking, "bodied" subject'; '2. Erasing or blurring of boundaries between inner … and outer … landscapes, or the erasing or blurring of self–other … distinctions'; '3. Re-eroticizing human relationships with a "bodied" landscape'; and also '7. Affirming the value of partial views and perspectives, the importance of "bioregions", and the locatedness of human subjects' (Legler 1997: 230–1), indicating the poet's departure from normative imaginings of the natural world.

15   Or in Eliotic terms, 'as if a magic lantern threw the nerves in patterns on a screen' (2015 [1917]: vol. I, 8, line 105).

16   The Faber *Poems* dates Bunting's ode as 1926; while Stevens's daughter recalls that 'The Idea of Order at Key West' first appeared in the quarterly *Alcestis* among 'a group of eight poems … in October 1934' (Stevens 1966: 256). It is unlikely that Stevens would have read Bunting's poem, though not impossible: Roger Guedalla notes in that *Redimiculum Matellarum*, the volume in which Bunting's ode first appeared, was 'Published March 1930, privately' at Milan,

and claimed in its opening pages to be 'copyright in all civilised countries but not (yet) in the United States' (Guedalla 1973: 13). Guedalla continues that 'The book's publication went totally unnoticed, except for a review by Louis Zukofsky' (ibid.: 14). Nevertheless, Guedalla records that in *Poetry* 27 (1) (October 1930), 'The "News Notes" p.59 refer to the ... publication of *Redimiculum Matellarum*' (ibid.: 75). Stevens was a sometime contributor to *Poetry*, so may have seen the notice, though his most recent publication there, 'Hibiscus on the Sleeping Shores', according to a search on poetryfoundation. org (accessed 6 November 2016), had been nine years earlier in *Poetry* 19 (1) (October 1921).

17 Even if we are to read the littoral in the Bunting's poem as libidinal – that is, not literal – his analogy still needs to imagine an uncontrollable, nonhuman tidal energy to operate.

Chapter 5: David Jones's *Anathemata* and the Gratuitous Environment

1 I refer to *The Anathemata* throughout this chapter as a 'work' rather than 'poem', in acknowledgement of its mixture of verse, prose, illustration, inscription and annotation.

2 Will Steffen, Paul J. Crutzen and John R. McNeill define the term, which they say 'suggests that the Earth has now left its natural geological epoch, the present interglacial state called the Holocene. Human activities have become so pervasive and profound that they rival the great forces of Nature and are pushing the Earth into planetary *terra incognita*. The Earth is rapidly moving into a less biologically diverse, less forested, much warmer, and probably wetter and stormier state' (2007: 614, authors' italics). However, while the term is still to be officially endorsed at the time of writing, Timothy Clark observes that 'its force is mainly as a loose, shorthand term for all the new contexts and demands – cultural, ethical, aesthetic, philosophical and political – of environmental issues that are truly planetary in scale, notably climate change' among others (2015: 2).

3 To relegate a species to a 'by-product' is to make it subsidiary to a particular process; compare Ulrich Beck's analysis of the 'side effect' in risk society (2009: 19) that I return to later in this chapter.

4 In terms that have an unintentional resonance with Dawson's Edenic idea of the period, Cherry Lewis notes that 'The Tertiary is frequently referred to by geologists as "the gardening on the top" since, to a geologist, the rocks are so

young – the Tertiary started 65 million years ago', ending with the advent of what is commonly referred to as the ice age (2000: 24).

5   Morton reminds us that 'Ideologies are commands pretending to be descriptions' (2010a: 131), and Diamond's 'description' of human development is inflected by his sense of progress.

6   While Jones writes before present understanding of climate change, the view he expresses in this passage is in keeping with mid-century speculation about another ice age rather than an anthropogenically warmed world. Trexler points out that 'In the 1970s, the rise of environmentalism led to more widespread concern with human impacts on global climate, although there was confusion in the mass media over whether melting ice caps or a new ice age were more likely' (2015: 3). Inasmuch, Jones is faithful to a 'permanent mythus' in which human existence is contingent on an enabling environment, even though he imagines it according to the scientific thinking of his day.

7   While Dawson and Diamond both do the same in their own ways, their ideologies are implicit in the structure of their respective accounts and see themselves according to an uninterrogated narrative of progress associated with science, rather than openly acknowledging the way they make sense of the historical and geological records.

8   For instance, the most recent report of the Intergovernmental Panel on Climate Change (IPCC) says that 'The evidence for human influence on the climate system has grown since the IPCC Fourth Assessment Report (AR4) [2007]. It is *extremely likely* that more than half of the observed increase in global average surface temperature from 1951 to 2010 was caused by the anthropogenic increase in GHG [greenhouse gas] concentrations and other anthropogenic forcings together' (2014: 5; authors' italics), where extremely likely = 95–100%.

9   This compares with the more common Biblically derived myths in which we accommodate our understanding of climate change according to Hulme (2009: 340–55).

10   This again speaks to the value of Paul Sheehan's assertion that narrative 'is *human-shaped*' (2002: 9; author's italics).

11   Remember it is the "reality" of trees that prompts the key, early reflection about the role of nature in literature in Buell's *The Environmental Imagination* (1995: 10).

12   As in the work of Beth Dempster, discussed in Chapter 4, and Naudts et al.'s study of European forestry practices in the last three centuries (2016; see note 7 to the previous chapter).

13   Jones also exercised this principle in his practice as a visual artist. He wrote to

René Hague, on 2 December 1935: 'How to make it not *realistic* is the bugger' (Jones 1980: 80; author's italics). This can be seen in the patterning of his 1931 drawing 'Merlin-land' (Jones 2010 [1952]; facing 185), which opens section VII of *The Anathemata*, 'Mabinog's Liturgy'. In the image, animals, trees and a human figure are organized loosely around a lower centre and their poses, whether in foreground or background, resemble one another, largely gesturing towards that centre. None of the subjects is rendered 'realistically', and the flattened effect gives a beast and a monument in the 'background' – a natural entity and a cultural one – as much clarity as those lower down it, as though in a fractal, vortex-like arrangement.

14   Crawford points out that for the first 'book-length version' of *The Waste Land*, Eliot 'added notes written during the summer' following the poem's composition; however, 'there was still too little material to fill even a small volume, so … he extended his notes in a way that he later regretted' (Crawford 2015: 423).

15   The lines, Jones explains in a note, were inspired by his 'maternal grandmother [who] was saddened by the [lavender-seller's] call, because she said it meant that summer was almost gone and that winter was again near' (2010 [1952]: 125 n.1).

16   Charles Tomlinson, for instance, criticizes 'the faults of over-reference' in *The Anathemata* (1983: 12): 'one remains uncomfortably aware that any given insight is likely to be crushed by imaginative over-crowding, by relentless typological parallels' (ibid.: 15).

17   Timothy Clark notes that: 'The original coiners of the term dated the Anthropocene from the industrial revolution and the invention of the steam engine' (2015: 1), though a number of possible timeframes have since been advanced by other scientists, from the beginnings of agriculture to the use of nuclear weapons (see Lewis and Maslin 2015).

18   Hague explains that Jones 'frequently … uses "he" or "his", "him" etc., to indicate that, while he has an individual in mind, that individual is to be regarded as typical' (Hague 1977: 11).

Chapter 6: The Poems of Our Climate Change

1   The title of Reading's collection, a reference to absolute zero, suggests the attempt to stabilize or counteract the process of global warming with which it engages.

2   The campaign aims to raise awareness of the holdings of major corporations,
     trusts and other bodies in the fossil fuel industry and to encourage their
     divestment.

3   'Parliament', an ecologically conscious recycling of Chaucer's *Parlement of
     Foules*, first appears in Duffy's collection *The Bees*, which contains a number of
     pieces dealing with climate change and environmental issues, such as 'Bees' and
     'The English Elms' (Duffy 2011: 50–1, 3, 40–1).

4   Jones's remarks are also echoed by Bunting in the preface to the 1968 edition
     of his *Collected Poems*, when he writes: 'I have set down words as a musician
     pricks his score, not to be read in silence, but to trace in the air a pattern of
     sound' (2016 [1968]: 555); this method is itself alluded to in *Briggflatts* with
     Bunting's reference to 'laying the tune frankly on the air' (2016 [1965]: 42; part
     I, line 75) or simply 'laying the tune on the air' (ibid.: 46; part II, line 76).

5   For a more detailed exploration of this idea, see Griffiths (2017).

6   While this 'his' could refer to the 'upstairs neighbor' mentioned earlier in 'Full
     Fathom' (Graham 2008: 30), a page elapses between his apparent abduction and
     the passage quoted above. In this case, then the poem also suggests a domestic
     complicity in neglecting the chain of environmental consequence Graham has
     traced. Despite the neighbour's arrest or rendition – 'they took him / away
     … how frightened you knew he was' – the 'you' who is in turn addressed or
     accused by the narrator 'did not / protect' him and 'went on with your / day'
     (ibid.).

7   As in Stevens's 'A Postcard from the Volcano', Graham's reference to the 'dead
     gods' suggests that we are still hidebound by superseded conceptions of the
     world, yet ignorant of this state.

8   Such a concentration into natural observation and bodily experience is a
     strategy that marks the conclusion of several of the poems in the collection;
     'Summer Solstice', for example, closes with the image of a dove alighting in an
     acacia, 'making its nest again this year … as if all time / came down to / this'
     (Graham 2008: 29). The conditional 'as if' here marks the fictive quality of the
     resolution.

9   Graham's setting is reminiscent of Ariel 'glad he had written his poems' in
     Stevens's 'The Planet on the Table' (1997 [1954]: 450), though the mood is
     substantially different. Jo Shapcott's 'Composition' makes an attempt comparable
     to Graham's to represent climate change as occurring in the context of writing:
     'the tea cups / wanted washing and the Gulf Stream / was slowing and O my
     hips // ached from sitting' (2010: 51). Shapcott's poem seeks a more lyrical than
     experimental reconciliation with the phenomena, however, in its emplaced

movement from its first lines, 'And I sat among the dust motes, my pencil /
(blue) sounding loud on the page,' to its last, 'and then there was this' (ibid.). In
the 'Acknowledgements' to her book, Shapcott thanks 'the neuro-scientist Mark
Lythgoe who, for the poem 'Composition', introduced me to latent inhibition
(the ability we have to filter out irrelevant stimuli)' (ibid.: n.p.), and the poem
associates the suspension of this facility with creative endeavour; while this
enables some productive disruptions of scale in her poem, there is no sense that
the slowing of the Gulf Stream, or indeed the collapsing ice shelf mentioned
later in 'Composition', remain anything other than extraneous, 'irrelevant', a
troubling context that is kept at a distance by the closure of the piece.

10   Remember that the invocation of the context of composition is not only a
feature of the opening sonnet of 'The Sorcerer's Mirror' but an ecocritical trope
exercised by Lawrence Buell in arguing for the value of the referential, and
(Buell 1995: 10) and later parodied by Timothy Morton (Morton 2007: 29).

11   Graham's 15,000-year time frame for this is at odds with the 10,000–12,000
years I have suggested earlier, though these are, unlike Motion's three millennia,
of a similar order of magnitude. This variation indicates the range of scientific
accounts of the duration of our window for civilization, without denying that
civilization depends on it.

12   Her spring also echoes the underground stirrings of *The Waste Land*, written in
'the shadow of war' itself.

References

Ackroyd, Peter (1985), *T. S. Eliot*, London: Abacus-Sphere.

Adcock, Fleur (2000), *Poems 1960–2000*, Newcastle: Bloodaxe.

Alaimo, Stacy (2010), *Bodily Natures: Science, Environment, and the Material Self*, Bloomington: Indiana University Press.

Alexander, Michael (1998 [1979]), *The Poetic Achievement of Ezra Pound*, Edinburgh: Edinburgh University Press.

Armitage, Simon (2010), 'The Present', *Guardian*, 16 October. Available online: www.theguardian.com/books/2010/oct/16/simon-armitage-present-prize-poem (accessed 28 April 2016).

Armstrong, Isobel (1982), *Language as Living Form in Nineteenth-Century Poetry*, Brighton: Harvester Press; Totowa, NJ: Barnes and Noble.

Armstrong, Tim (2005), *Modernism: A Cultural History*, Cambridge: Polity.

Astley, Neil, ed. (2007), *Earth Shattering: Ecopoems*, Tarset: Bloodaxe.

Bate, Jonathan (1991), *Romantic Ecology: Wordsworth and the Environmental Tradition*, London: Routledge.

Bate, Jonathan (1996), 'Living with the Weather', *Studies in Romanticism* 35 (3): 431–47.

Bate, Jonathan (2000), *The Song of the Earth*, London: Picador.

Beck, Ulrich (2009), *World at Risk*, trans. Ciaran Cronin, Cambridge: Polity.

Behringer, Wolfgang (2010), *A Cultural History of Climate*, trans. Patrick Camiller, Cambridge: Polity.

Bennett, Jane (2010), *Vibrant Matter: A Political Ecology of Things*, Durham, NC: Duke University Press.

Botkin, Daniel B. (2012), *The Moon in the Nautilus Shell: Discordant Harmonies Reconsidered*, Oxford: Oxford University Press.

Box, George E. P. (1979), 'Robustness in the Strategy of Scientific Model Building', in R. L. Launer and G. N. Wilkinson (eds), *Robustness in Statistics*, 201–36, New York: Academic Press.

Brown, Phillip (2001), 'A Northern Lucretius: Watching Things Work out Their Own Fate in *Briggflatts*', [Typescript] Basil Bunting Poetry Archive, Palace Green Library, Durham University, Durham.

Buell, Frederick (2004), *From Apocalypse to Way of Life: Environmental Crisis in the American Century*, New York: Routledge.

Buell, Lawrence (1995), *The Environmental Imagination: Thoreau, Nature Writing, and the Formation of American Culture*, Cambridge, MA: Belknap-Harvard University Press.

Buell, Lawrence (2001), *Writing for an Endangered World: Literature, Culture, and Environment in the U.S. and Beyond*, Cambridge, MA: Belknap-Harvard University Press.

Bunting, Basil (1932), [Typescript] letter to Harriet Monroe, 20 November, Box 1, Folder 12, Morton Dauwen Zabel Papers, Special Collections Research Center, University of Chicago Library.

Bunting, Basil (1933), [Typescript] letter to Morton Dauwen Zabel, 4 January, Box 1, Folder 12, Morton Dauwen Zabel Papers, Special Collections Research Center, University of Chicago Library.

Bunting, Basil (1978), 'Basil Bunting Talks About *Briggflatts*', interview by Peter Quartermain and Warren Tallman, *Agenda* 16 (1): 8–19.

Bunting, Basil (2009), *Briggflatts*, Tarset: Bloodaxe.

Bunting, Basil (2016), *The Poems of Basil Bunting*, ed. Don Share, London: Faber and Faber.

Burton, Richard (2013), *A Strong Song Tows Us: The Life of Basil Bunting*, Oxford: Infinite Ideas.

Cantrell, Carol H. (2003 [1998]), '"The Locus of Compossibility": Virginia Woolf, Modernism, and Place', in Michael P. Branch and Scott Slovic (eds), *The ISLE Reader: Ecocriticism, 1993–2003*, 33–48, Athens, GA: University of Georgia Press.

Chasseaud, Peter (1997), 'David Jones and the Survey', in Paul Hills (ed.), *David Jones, Artist and Poet*, 18–30, Aldershot: Scolar Press.

Clark, Nigel (2010), 'Volatile Worlds, Vulnerable Bodies: Confronting Abrupt Climate Change', *Theory, Culture and Society* 27 (2–3): 31–53.

Clark, Nigel (2012), 'Rock, Life, Fire: Speculative Geophysics and the Anthropocene', *Oxford Literary Review* 34 (2): 259–76.

Clark, Timothy (2011), *The Cambridge Introduction to Literature and the Environment*, Cambridge: Cambridge University Press.

Clark, Timothy (2012), 'Scale: Derangements of Scale', in Tom Cohen (ed.), *Telemorphosis: Theory in the Era of Climate Change*, vol. I, 149–67, Ann Arbor, MI: Open Humanities Press.

Clark, Timothy (2013), 'What on World is the Earth? The Anthropocene and Fictions of the World', *Oxford Literary Review* 35 (1): 5–24.

Clark, Timothy (2015), *Ecocriticism on the Edge: The Anthropocene as a Threshold Concept*, London: Bloomsbury.

Conniff, Brian (1988), *The Lyric and Modern Poetry: Olson, Creeley, Bunting*, New York: Lang.

Coomber, John (2012), 'Moving Beyond the Uncertainty of Climate Change Risk: Applying the Insurance Principles of Measurement, Mitigation and Diversification to the World's Most Challenging Risk', London, ClimateWise. Available online: http://static1.1.sqspcdn.com/static/f/270724/18214876/1337098418780/ClimateWise+Thought+Leadership+Series+2012+-+Issue+One+John+Coomber.pdf?token=iCz7bpU2ODWOCYh%2BWFx7%2BbIc8k4%3D. Web. (accessed 28 April 2016).

Costello, Bonnie (1998), '"What to Make of a Diminished Thing": Modern Nature and Poetic Response', *American Literary History* 10 (4): 569–605.

Costello, Bonnie (2007), 'US Modernism I: Moore, Stevens and the Modernist Lyric', in Alex Davis and Lee M. Jenkins (eds), *The Cambridge Companion to Modernist Poetry*, 163–80, Cambridge: Cambridge University Press.

Crawford, Robert (1987), *The Savage and the City in the Work of T. S. Eliot*, Oxford: Oxford University Press-Clarendon.

Crawford, Robert (2015), *Young Eliot: From St Louis to* The Waste Land, London: Jonathan Cape.

Dawson, Christopher (1933 [1928]), *The Age of the Gods: A Study in the Origins of Culture in Prehistoric Europe and the Ancient East*, London: Sheed and Ward.

The Day After Tomorrow (2004), [Film] Dir. Roland Emmerich, USA: Fox.

Dempster, Beth (2007), 'Boundarylessness: Introducing a Systems Heuristic for Conceptualising Complexity', in Charles S. Brown and Ted Toadvine (eds), *Nature's Edge: Boundary Explorations in Ecological Theory and Practice*, 93–108, Albany: State University of New York Press.

Diamond, Jared (1998), *Guns, Germs and Steel: A Short History of Everybody for the Last 13,000 Years*, London: Vintage.

Dilworth, Thomas (2008), *Reading David Jones*, Cardiff: University of Wales Press.

Duffy, Carol Ann (2011), *The Bees*, London: Picador-Macmillan.

Eliot, T. S. (1957), *On Poetry and Poets*, London: Faber and Faber.

Eliot, T. S. (1975), *Selected Prose of T. S. Eliot*, ed. Frank Kermode, London: Faber and Faber.

Eliot, T. S. (2015), *The Poems of T. S. Eliot*, eds Christopher Ricks and Jim McCue, 2 vols, London: Faber and Faber.

Ellis, Steve (1993), *West Pathway: New Poems*, Newcastle: Bloodaxe.

Ellmann, Maud (1987), *The Poetics of Impersonality: T. S. Eliot and Ezra Pound*, Cambridge, MA: Harvard University Press.

Felstiner, John (2009), *Can Poetry Save the Earth? A Field Guide to Nature Poems*, New Haven, CT: Yale University Press.

Flannery, Tim (2006), *We are the Weather Makers: The Story of Global Warming*, London: Penguin.

Fletcher, Angus (2004), *A New Theory for American Poetry: Democracy, the Environment and the Future of Imagination*, Cambridge, MA: Harvard University Press.

Free Word Centre (2015), 'Weatherfronts: Climate Change and the Stories We Tell', London: Tipping Point/Free Word Centre. Available online: www.freewordcentre. com/assets/public/files/Weathfronts_Commissions_Publication_2015.pdf (accessed 29 April 2016).

Gibbon, Edward (1994 [1781–8]), *The History of the Decline and Fall of the Roman Empire*, ed. David Womersley, 3 vols, London: Penguin.

Graham, Jorie (2008), *Sea Change: Poems*, Manchester: Carcanet.

Greaves, Sara R. (2005), 'A Poetics of Dwelling in Basil Bunting's *Briggflatts*', *Cercles* 12: 64–78.

Greenlaw, Lavinia (1993), *Night Photograph*, London: Faber and Faber.

Grey, Thomas C. (1991), *The Wallace Stevens Case: Law and the Practice of Poetry*, Cambridge, MA: Harvard University Press.

Griffiths, Matthew (2012), 'Tensions in the Mesh: Thoughts on *The Ecological Thought*', review of *The Ecological Thought* by Timothy Morton, *Oxford Literary Review* 34 (2): 326–31.

Griffiths, Matthew (2014), 'Climate Change and the Individual Talent: Eliotic Ecopoetics', *symplokē* 21 (1–2): 83–95.

Griffiths, Matthew (2017), 'Jorie Graham's *Sea Change*: The Poetics of Sustainability and the Politics of what we're Sustaining', in Adeline Johns-Putra, John Parham and Louise Squire (eds), *Literature and Sustainability: Exploratory Essays*, Manchester: Manchester University Press.

Guardian, The (2015), '"Our melting, shifting, liquid world": Celebrities read poems on climate change'. Available online: www.theguardian.com/environment/ ng-interactive/2015/nov/20/our-melting-shifting-liquid-world-celebrities-read-poems-on-climate-change (accessed 29 April 2016).

Guedalla, Roger (1973), *Basil Bunting: A Bibliography of Works and Criticism*, Norwood, PA: Norwood Editions.

Habib, M. A. R. (1999), *The Early T. S. Eliot and Western Philosophy*, Cambridge: Cambridge University Press.

Hague, René (1977), *A Commentary on* The Anathemata *of David Jones*, Wellingborough: Skelton.

Harrison, Robert Pogue (1993), *Forests: The Shadow of Civilization*, Chicago: University of Chicago Press.

Harrison, Robert Pogue (1999), 'Not Ideas about the Thing but the Thing Itself', *New Literary History* 30 (3): 661–73.

Hatlen, Burton (2000), 'Regionalism and Internationalism in Basil Bunting's *Briggflatts*', *Yale Journal of Criticism* 13 (1): 49–66.

Heaney, Seamus (2006), *District and Circle*, London: Faber and Faber.

Howarth, Peter (2012), *The Cambridge Introduction to Modernist Poetry*, Cambridge: Cambridge University Press.

Hulme, Mike (2009), *Why We Disagree About Climate Change: Understanding Controversy, Inaction and Opportunity*, Cambridge: Cambridge University Press.

An Inconvenient Truth (2006), [Film] Dir. Davis Guggenheim, USA: Paramount and UIP.

Intergovernmental Panel on Climate Change (2008), 'Climate Change 2007: Synthesis Report', Geneva: Intergovernmental Panel on Climate Change. Available online: www.ipcc.ch/pdf/assessment-report/ar4/syr/ar4_syr.pdf (accessed 28 April 2016).

Intergovernmental Panel on Climate Change (2015), 'Climate Change 2014: Synthesis Report', Geneva: Intergovernmental Panel on Climate Change. Available online: http://ipcc.ch/pdf/assessment-report/ar5/syr/SYR_AR5_FINAL_full_wcover.pdf (accessed 28 April 2016).

Iovino, Serenella and Serpil Oppermann (2012), 'Theorizing Material Ecocriticism: A Diptych', *Interdisciplinary Studies in Literature and the Environment* 19 (3): 448–75.

Jones, David (1959), *Epoch and Artist: Selected Writings*, London: Faber and Faber.

Jones, David (1980), *Dai Greatcoat: A Self-portrait of David Jones in his Letters*, ed. René Hague, London: Faber and Faber.

Jones, David (1992), *Selected Works of David Jones from* In Parenthesis, The Anathemata, The Sleeping Lord, ed. John Matthias, Orono: National Poetry Foundation; Cardiff: University of Wales Press.

Jones, David (2010 [1952]), *The Anathemata*, London: Faber and Faber.

Kainulainen, Maggie (2014), 'Saying Climate Change: Ethics of the Sublime and the Problem of Representation', *symplokē* 21 (1–2): 109–23.

Kato, Daniela (2014), '"Distilled Essence of Cormorant": The Ecopoetics of *Briggflatts* and the Modernist Biomorphic Imagination', *Green Letters: Studies in Ecocriticism*, 18 (2): 154–69.

Keats, John (2001), *The Major Works*, ed. Elizabeth Cook, Oxford: Oxford University Press, 2001.

Kerridge, Richard (2007), 'Climate Change and Contemporary Modernist Poetry', in Tony Lopez and Anthony Caleshu (eds), *Poetry and Public Language*, 131–48, Exeter: Shearsman.

Kerridge, Richard (2013), 'Ecocriticism', *The Year's Work in Critical and Cultural Theory* 21 (1): 345–74.

Knickerbocker, Scott (2012), *Ecopoetics: The Language of Nature, the Nature of Language*, Amherst and Boston: University of Massachusetts Press.

Kolbert, Elizabeth (2007), *Field Notes from a Catastrophe: A Frontline Report on Climate Change*, London: Bloomsbury.

Lahr, Oonagh (1972), 'Greek Sources of "Writ in Water"', *Keats–Shelley Journal* 21–2: 17–18.

Lanchester, John (2007), 'Warmer, Warmer', review of various titles, *London Review of Books*, 22 March: 3–9.

Latour, Bruno (1993), *We Have Never Been Modern*, trans. Catherine Porter, New York: Harvester.

Latour, Bruno (2004), *Politics of Nature: How to Bring the Sciences into Democracy*, trans. Catherine Porter, Cambridge, MA: Harvard University Press.

Latour, Bruno (2006), *Reassembling the Social: An Introduction to Actor-Network-Theory*, Oxford: Oxford University Press.

Legler, Gretchen T. (1997), 'Ecofeminist Literary Criticism', in Karen J. Warren (ed.), *Ecofeminism: Women, Culture, Nature*, 227–38, Bloomington: Indiana University Press.

Lekan, Thomas M. (2014), 'Fractal Eaarth: Visualizing the Global Environment in the Anthropocene', *Environmental Humanities* 5: 171–201.

Lentricchia, Frank (1994), *Modernist Quartet*, Cambridge: Cambridge University Press.

Lewis, Cherry (2000), *The Dating Game: One Man's Search for the Age of the Earth*, Cambridge: Cambridge University Press.

Lewis, Simon L. and Mark A. Maslin (2015), 'Defining the Anthropocene', *Nature* 519: 171–80.

McCarthy, Jeffrey Mathes (2015), *Green Modernism: Nature & The English Novel 1900 to 1930*, New York: Palgrave Macmillan (ebook).

McDaniel, Judith (1974), 'Wallace Stevens and the Scientific Imagination', *Contemporary Literature* 15 (2): 221–37.

McGuffie, Kendal and Ann Henderson-Sellers (2005), *A Climate Modelling Primer*, 3rd edn, Chichester: Wiley.

McQuillan, Martin (2012), 'Notes Toward a Post-Carbon Philosophy: "It's the Economy, Stupid"', in Tom Cohen (ed.), *Telemorphosis: Theory in the Era of Climate Change*, vol. I, 270–92, Ann Arbor, MI: Open Humanities Press.

Markley, Robert (2012), 'Time: Time, History, and Sustainability' in Tom Cohen (ed.), *Telemorphosis: Theory in the Era of Climate Change*, vol. I, 43–64, Ann Arbor, MI: Open Humanities Press.

Marvell, Andrew (2005), *The Complete Poems*, ed. Elizabeth Story Donno, introduced by Jonathan Bate, rev. edn, London: Penguin.

Mellors, Anthony (2005), *Late Modernist Poetics from Pound to Prynne*, Manchester: Manchester University Press.

Miller, J. Hillis (2010), 'Anachronistic Reading', *Derrida Today* 3 (1): 75–91.

Morton, Timothy (2007), *Ecology without Nature: Rethinking Environmental Aesthetics*, Cambridge, MA: Harvard University Press.

Morton, Timothy (2010a), *The Ecological Thought*, Cambridge, MA: Harvard University Press.

Morton, Timothy (2010b), 'Ecology as Text, Text as Ecology', *Oxford Literary Review* 32 (1): 1–17.

Motion, Andrew (1997), *Keats*, London: Faber and Faber.

Motion, Andrew (2009a), 'Andrew Motion Warms to Poem About Climate Change', interview by Richard Eden, *Telegraph*, 2 May. Available online: www.telegraph.co.uk/news/newstopics/mandrake/5263156/Andrew-Motion-warms-to-poem-about-climate-change.html (accessed 29 April 2016).

Motion, Andrew (2009b), 'The Sorcerer's Mirror', *Guardian*, 26 September. Available online: www.theguardian.com/culture/2009/sep/26/the-sorcers-mirror-andrew-motion (accessed 29 April 2016).

Murphy, Patrick D. (2000), *Farther Afield in the Study of Nature-Oriented Literature*, Charlottesville: University Press of Virginia.

Murphy, Patrick D. (2015), *Persuasive Aesthetic Ecocritical Praxis: Climate Change, Subsistence, and Questionable Futures*, Lanham, Boulder, New York and London: Lexington.

Murray, Les (1989), 'The Greenhouse Vanity', *London Review of Books*, 18 May: 8.

Murray, Les (2003), *New Collected Poems*, Manchester: Carcanet.

Naudts, Kim, Yiying Chen, Matthew J. McGrath, James Ryder, Aude Valade, Juliane Otto and Sebastiaan Luyssaert (2016), 'Europe's Forest Management did not Mitigate Climate Warming', *Science* 351 (6,273): 597–600. Available online: http://science.sciencemag.org/content/351/6273/597 (accessed 29 April 2016).

Nicholls, Peter (2009), *Modernisms: A Literary Guide*, 2nd edn, Basingstoke: Macmillan.

North, Michael (2013), 'The Making of "Make it New"', *Guernica*, 15 August. Available online: www.guernicamag.com/features/the-making-of-making-it-new/ (accessed 29 April 2016).

Petrucci, Mario (2009), interview for *The Grove* by Matthew Griffiths, 31 October. Available online: www.mariopetrucci.com/interview-Grove.rtf (accessed 29 April 2016).

Pound, Ezra (2009 [1918]), 'A Retrospect'. Available online: www.poetryfoundation.org/learning/essay/237886 (accessed 29 April 2016).

Pound, Ezra (2011), *Selected Poems and Translations*, ed. Richard Sieburth, London: Faber and Faber.

Powell, D. A. (2009), *Chronic: Poems*, St Paul, MN: Graywolf.

Presley, Frances (2009), *Lines of Sight*, Exeter: Shearsman.

Rae, Simon (1991), *Soft Targets: From the* Weekend Guardian, illustrated by Willie Rushton, Newcastle: Bloodaxe.

Raine, Anne (2014), 'Ecocriticism and Modernism', in Greg Garrard (ed.), *The Oxford Handbook of Ecocriticism*, 98–117, Oxford: Oxford University Press.

Reading, Peter (2005), *–273.15*, Newcastle: Bloodaxe.

Roszak, Theodore (1995), 'Where Psyche Meets Gaia', in Theodore Roszak, Mary E. Gomes and Allen D. Kanner (eds), *Ecopsychology: Restoring the Earth, Healing the Mind*, 1–17, San Francisco: Sierra Club.

Royal Society of Arts (2015), *9 Original Poems on Climate Change*, London: Royal Society of Arts. Available online: https://www.thersa.org/globalassets/pdfs/events/climate-change-poetry-anthology.pdf (accessed 29 April 2016).

Sandars, N. K. (1976), 'The Present Past in *The Anathemata* and Roman Poems', in Roland Mathias (ed.), *David Jones: Eight Essays on His Work as Writer and Artist*, 50–72, Llandysul: Gomer.

Schuster, Joshua (2015), *The Ecology of Modernism: American Environments and Avant-Garde Poetics*. Tuscaloosa: University of Alabama Press.

Schwartz, Sanford (1985), *The Matrix of Modernism: Pound, Eliot, and Early Twentieth-Century Thought*, Princeton: Princeton University Press.

Scigaj, Leonard M. (1999), *Sustainable Poetry: Four American Ecopoets*. Lexington: University Press of Kentucky.

Sellars, Roy (2010), 'Waste and Welter: Derrida's Environment', *Oxford Literary Review* 32 (1): 37–49.

Shakespeare, William (1984), *Macbeth*, ed. Kenneth Muir, reprinted with new introduction, London: Methuen.

Shakespeare, William (1995), *Antony and Cleopatra*, ed. John Wilders, London: Routledge.

Shakespeare, William (2006), *As You Like It*, ed. Juliet Dusinberre, London: Thomson.

Shakespeare, William (2011), *The Tempest*, eds Virginia Mason Vaughan and Alden T. Vaughan, revised edn, London: Routledge.

Shakespeare, William and John Fletcher (2000), *King Henry VIII*, ed. Gordon McMullan. London: Thomson.

Shapcott, Jo (2010), *Of Mutability*, London: Faber and Faber.

Sharpe, Tony (2000), *Wallace Stevens: A Literary Life,* Basingstoke: Macmillan.

Sheehan, Paul (2002), *Modernism, Narrative and Humanism*, Cambridge: Cambridge University Press.

Shelley, Percy Bysshe (1977), *Shelley's Poetry and Prose: Authoritative Texts, Criticism*, selected and edited by Donald H. Reiman and Sharon B. Powers, New York: Norton.

Smith, Grover (1983), *The Waste Land*, London: Allen & Unwin.

Smith, Laurie (2008), 'The New Imagination', *Magma* 42: 16–22.

Solnick, Sam (2016), *Poetry and the Anthropocene: Ecology, Biology and Technology in Contemporary British and Irish Poetry*, Abingdon: Routledge.

Soper, Kate (1995), *What Is Nature? Culture, Politics and the Non-Human*, Oxford: Blackwell.

Soper, Kate (2011), 'Passing Glories and Romantic Retrievals: Avant-Garde Nostalgia and Hedonist Renewal', in Axel Goodbody and Kate Rigby (eds), *Ecocritical Theory: New European Approaches*, 17–29, Charlottesville: University of Virginia Press.

Stanbridge, Paul (2011), 'The Making of David Jones's *Anathemata*', PhD diss., University of East Anglia, Norwich.

Steffen, Will, Paul J. Crutzen and John R. McNeill (2007), 'The Anthropocene: Are Humans Now Overwhelming the Great Forces of Nature?', *Ambio* 36 (8): 614–21.

Stern, Nicholas (2008), *The Economics of Climate Change: The Stern Review*, 4th printing, Cambridge: Cambridge University Press.

Stevens, Holly (1971), 'Bits of Remembered Time', *Southern Review* 7 (3): 651–7.

Stevens, Wallace (1921), 'Hibiscus on the Sleeping Shores', *Poetry* (October): 9.

Stevens, Wallace (1966), *Letters of Wallace Stevens*, selected and edited by Holly Stevens, London: Faber and Faber.

Stevens, Wallace (1997), *Collected Poetry & Prose*, selected and annotated by Frank Kermode and Joan Richardson, New York: Library of America.

Sullivan, Heather I. (2012), 'Dirt Theory and Material Ecocriticism', *Interdisciplinary Studies in Literature and the Environment* 19 (3): 516–31.

Summerfield, Henry (1979), *An Introductory Guide to* The Anathemata *and* The Sleeping Lord *Sequence of David Jones*, Victoria, BC: Sono Nis Press.

Swyngedouw, Erik (2010), 'Apocalypse Forever? Post-Political Populism and the Spectre of Climate Change', *Theory, Culture & Society* 27 (2–3): 213–32.

Szerszynski, Bronislaw (2007), 'The Post-Ecologist Condition: Irony as Symptom and Cure', *Environmental Politics* 16 (2): 337–55.

Szerszynski, Bronislaw (2010), 'Reading and Writing the Weather: Climate Technics and the Moment of Responsibility', *Theory, Culture & Society* 27 (2–3): 9–30.

Tarlo, Harriet (2000), 'Radical Landscapes: Contemporary Poetry in the Bunting

Tradition', in James McGonigal and Richard Price (eds), *The Star You Steer By: Basil Bunting and British Modernism*, 149–80, Amsterdam: Rodopi.

Taylor, Jesse Oak (2013), 'The Novel as Climate Model: Reading the Greenhouse Effect in *Bleak House*', *Novel: A Forum on Fiction* 46 (1): 1–25.

Thomas, M. Wynn (2008), 'Full Fathom Five', review of *Sea Change* by Jorie Graham, *Guardian*, 3 May. Available online: www.theguardian.com/books/2008/may/03/featuresreviews.guardianreview25 (accessed 29 April 2016).

Thormählen, Marianne (1978), *The Waste Land: A Fragmentary Wholeness*, Lund: CWK Gleerup Lund.

Tomlinson, Charles (1983), *The Sense of the Past: Three Twentieth-Century British Poets*, Liverpool: University of Liverpool Press.

Trexler, Adam (2011), 'Economics', in Jason Harding (ed.), *T. S. Eliot in Context*, 275–84, Cambridge; Cambridge University Press.

Trexler, Adam (2015), *Anthropocene Fictions: The Novel in a Time of Climate Change*, Charlottesville and London: University of Virginia Press.

Tuana, Nancy (2008), 'Viscous Porosity: Witnessing Katrina', in Stacy Alaimo and Susan Hekman (eds), *Material Feminisms*, 188–213, Bloomington: Indiana University Press.

Vendler, Helen (1995), *The Given and the Made: Recent American Poets*, London: Faber and Faber.

Vendler, Helen (1996 [1984]), *Wallace Stevens: Words Chosen Out of Desire*, 4th printing, Cambridge, MA: Harvard University Press.

Voros, Gyorgyi (1997), *Notations of the Wild: Ecology in the Poetry of Wallace Stevens*, Iowa City: University of Iowa Press.

Waugh, Patricia (1992), *Practising Postmodernism, Reading Modernism*, London: Arnold.

Waugh, Patricia (2001), *Beyond Mind and Matter: Scientific Epistemologies and Modernist Aesthetics*, Aalborg: Significant Forms: The Rhetoric of Modernism.

Whitmarsh, Lorraine (2009), 'What's in a Name? Commonalities and Differences in Public Understanding of "Climate Change" and "Global Warming"', *Public Understanding of Science* 18 (4): 401–20.

Williams, William Carlos (2000), *Collected Poems*, eds. A. Walton Litz and Christopher MacGowan, 2 vols, Manchester: Carcanet.

Wood, David (2005), *The Step Back: Ethics and Politics after Deconstruction*, Albany: State University of New York Press.

Woodman, A. J. (1975), 'Greek Sources of "Writ in Water": A Further Note', *Keats–Shelley Journal* 24: 12–13.

Wordsworth, William (1979), *The Prelude: 1799, 1805, 1850: Authoritative Texts, Context and Reception, Recent Critical Essays*, eds. Jonathan Wordsworth, M. H. Abrams and Stephen Gill, New York: Norton.

Worster, Donald (1985), *Nature's Economy: A History of Ecological Ideas*, new edn, Cambridge: Cambridge University Press.

Index

Lightning Source UK Ltd.
Milton Keynes UK
UKHW021048021219
354618UK00003B/268/P